Readings in Environmental Ethics

Readings in Environmental Ethics

Multidisciplinary Perspectives

Edited by
D.C. *Srivastava*

RAWAT PUBLICATIONS
Jaipur and New Delhi

ISBN 81-7033-893-X

Published by
Prem Rawat for *Rawat Publications*
Satyam Apts., Sector 3, Jawahar Nagar, Jaipur 302 004 (India)
Phone: 0141 265 1748 / 7006 Fax: 0141 265 1748
E-mail: info@rawatbooks.com
Website: www.rawatbooks.com

Delhi Office
4858/24, Ansari Road, Daryaganj, New Delhi 110 002
Phone: 011-23263290

Typeset by Rawat Computers, Jaipur
Printed at Chaman Enterprises, New Delhi

For
My Daughters

Ishita and *Smita*

in the hope that they may live
deeply valuable lives in this
eco-centric moral world

Contents

IV
Science, Society and Environment

Acknowledgements

I express my indebtedness and gratitude to Professor Bijoy H. Boruah, Professor of Philosophy, IIT, Kanpur, for the sustained interest he has taken in motivating and guiding me from time to time to produce this volume. Without his help and guidance I would not have been able to undertake this task.

I owe a special debt of gratitude to Professor S.A. Shaida, who has been my teacher and my guide. He has been a constant source of inspiration and encouragement.

My utmost gratitude goes, of course, to the contributors to this volume, without whose cooperation this project could not have been executed at all.

I have received help in the editorial work from some of my colleagues of Christ Church College, Kanpur, especially from Dr. (Mrs.) Madhumita Gangully, who has graciously offered valuable suggestions.

I would like to express heartfelt gratitude to Dr. Pervez E. Deen, the Principal of Christ Church College, Kanpur, who has been a source of inspiration and encouragement in undertaking this project. It was due to the blessings of Dr. Deen that I could delve into the problems of environmental ethics.

I am thankful to my colleagues and friends, especially Dr. Ashutosh Saxena, Dr. Sujata Chaturvedi, Dr. A.K.Singh, Dr. Ranjay Pratap Singh and Dr. Sanjay Kumar for making valuable suggestions at the various stages of the production of this volume.

I sincerely appreciate the indulgence shown to me by the members of my family during my preoccupation with this work.

I express my sincere most indebtedness to Dr. S.C. Srivastava, my elder brother, for his sustained help and encouragement in all walks of my life.

Last but not the least, I desire to thank the Rawat Publications for undertaking the task of publishing and printing this volume.

D.C. Srivastava

Introduction

Current thoughts on the issue of environment are keyed to a philosophical point, namely, a proper ethical characterization of the question of environment. For, at the very foundation of reflective awareness of the significance of the concept of environment, there lies the ethical essence of the concept. In other words, environmental awareness, as a serious mode of apprehending the significance of the environment against the background of the cosmic existence, is the awareness of environmental ethics.

With this foundational ethicality about environmental awareness in mind, reflections on the concept of environment take off in diverse directions. Of course, the diverse directions all have their ultimate anchorage in what might be characterized as "eco-centricity". The justification for this phrase rests on the fact that contemporary thoughts on environmentalism are triggered by a keen consciousness of lack of an eco-centric disposition. It would be quite plausible to say that contemporary deliberations of environmental ethics are motivated by a sustained and an unflinching search for environmental wisdom. The lack of an adequate eco-centric disposition is an expression of the lack of environmental wisdom.

Indeed, environmental wisdom is considered, whether tacitly or expressly, to be a central item in the agenda of contemporary humanity. Any exception to the practice of an environmentally wise course of action, collective or individual, is viewed as part of the process leading to a disastrous end to our survival. Hence, the imperative of environmental wisdom acts as the principal guiding force behind the various, more or less independent, discussions on the diverse aspects of environmentalism.

Readings in Environmental Ethics has been designed as an introduction to the central issues of environmentalism in general and environmental ethics in particular. This volume has been designed for readers who wish to explore in depth questions regarding *how on our planet earth we ought to live*, or as how we can live responsibly with the non-humans and the planet itself. Although primarily oriented towards questions and philosophical issues related to environmental ethics, this volume crosses academic disciplines and attends seriously to unearth and to extract critical attention, the relevant moral or normative presuppositions that underlie policy recommendations whether they are made for example by physicists, biologists, theologians, sociologists, economists or philosophers.

Given the status of the principal motivational basis of environmental research and deliberation, the question of environmental wisdom naturally figures as the basic thrust of any work on environmentalism. And this thrust is essentially ethical. Hence the present volume takes up the ethical issue of *Eco-centric Morality* as the opening section of the entire book. Contributors to this section grapple with the question of redefining the scope of ethics in such a way that the redefinition is able to accommodate the constraints of an environmental ethic. Can the concept of ethics be so redefined as to go beyond human-centred morality?

The philosophical section of the book takes up the topic of eco-centric morality and grapples with the question of whether a teleological interpretation of ecological ethics would be compatible with the anthropocentrism that is implicit in the usual conception of morality. The discussion centres on the competing claims and arguments between the anthropocentric and cosmocentric view of ethics, and then tilts towards a cosmocentric approach as theoretically more suitable framework for accommodating the overriding ideal of ecological balance.

The first essay by Professor S.A. Shaida has brought to the fore the historical and conceptual background of Environmental Ethics. Aristotle's idea of ethics as a practical science is held to be a central forerunner of the idea of applied ethics. The essay focuses on the turn away from Kantian apriorism and the advocacy of virtue-theoretic ethics that lend support to the general approach of applied ethics. This new turn is considered to be an Aristotelian preference of practical virtues underlying the concept of good life. As a fitting concern with the practical turn, Dr. Shaida emphasized on specific moral problems such as ecological degradation and disequilibrium. The point emphatically made by him is that an attitudinal paradigm shift from anthropocentricity to eco-centricity is most desirable in the present crisis of humanity.

Professor Rajendra Prasad in his essay entitled *On the Adequacy of the Human-centric Model of the Field of Moral Relations* argues for the adequacy of human-centric ethics for providing room for an environmental ethic on the ground that human-centric ethics can be rationally extended for such a purpose. The extension of ethical scope is predicated on the attitude of "humanization" of nature as a whole, meaning thereby that non-human entities can be regarded as "human-like" and therefore considered, in an extended sense, as deserving ethical "subjects". However, Prasad insists "there is no escape from accepting the primacy of the human, or from adopting the human-centric point of view".

Professor Bijoy H. Boruah in his essay entitled *Environmental Wisdom* emphasizes on the principle of interdependence as providing the needed theoretical underpinning of environmental ethics. Deep ecology which epitomizes environmental wisdom is said to be founded upon the fact that everything is interdependent in reality. It is a sincere recognition of the truth of this underlying principle that ought to motivate humanity to adopt a cosmocentric outlook in its dealing with the larger environment. Dr. Boruah argues that the Buddhist metaphysics of 'emptiness' with its core doctrine of 'Dependent Origination' is uniquely relevant and attuned to the environmental ethics that draws upon Deep Ecology. A proper environmental ethics that transcends ethical anthropocentricity logically takes its course towards ethical cosmocentric wisdom.

Dr. S.K. Pal in his essay entitled *Eco-centrism Revisited* contends that total egalitarianism of eco-centrism is neither possible nor deplorable, and that a balanced view of eco-centric ethics can be held only if the fact/value dichotomy is overcome. His point is that some sort of priority must be accorded to humans on the matter of value distribution, because humans are the creator and the evaluator of values. While blind anthropocentrism has deplorable consequences for the non-human world, blindly misanthropic eco-centrism has no less deplorable consequences to the human world. There is, according to Pal, a significant asymmetry between humans and non-humans, which does not call for any chauvinistic warning. The asymmetry is that, unlike the non-humans, the humans are the originator and evaluator of values and have the privilege of theorizing ethically about all existence.

Pal discerns a subtle dualism of fact and value even in Deep Ecology, which he otherwise endorses from his eco-centric perspective.

Deep Ecology does not fully overcome the duality of humans and nature. Pal at last refers to the Indian Vedantic metaphysical world-view as fully undercutting the fact/value distinction and providing a balanced eco-centric ethics.

Dr. Ramdas Sirkar in his essay advocates Peter Singer's extension of the boundary of ethics beyond the human species in order to include all sentient creatures having interests. But this extension does not mean that traditional anthropocentric ethics rules out the possibility of incorporating environmental values.

Dr. B.P. Patra in his essay entitled *Does Anthropocentric View Rest on a Mistake?* pleads for a cosmocentric metaphysical basis for an ideal moral theory that can adequately incorporate an environmental ethic. Criticizing anthropocentric morality that dominates modern western thought, he advocates the recent western turn towards deep ecology, but insists that deep ecology needs its metaphysical foundation, which is provided by traditional Indian Samkhya philosophy and the Upanishadic worldview.

Dr. A. Raghuramraju in the essay entitled *Tracing the (Dis) continuities: Heidegger on Technology* attributes the cause of ecological devastation in the wake of modern technological culture to the mindset of the modern theories of man that germinated in the philosophical ideas of the so-called Social Contract Theories of Hobbes, Locke and Rousseau. The pre-modern man is said to be "natural", whereas the image of the modern man is that of an individual "free" from nature and also free to dominate nature.

The essays by Professor R.C. Sinha and Dr. D.C. Srivastava grapple with the question of redefining the scope of ethics in such a way that the redefinition is able to accommodate the constraints of an environmental ethic. Both the essays focus upon the discussion centring on the competing claims and arguments between the anthropocentric and cosmocentric views of ethics. It has been emphasized that cosmocentric approach is theoretically more suitable framework for accommodating the overriding ideal of ecological balance.

Section II of the volume is *Religion, Culture and Environment*. This section deals with various religious-cultural perspectival outlooks on the relation between nature and human beings. The opening essay is on Christian perspective by Rev. Richard Howell. Richard Howell highlights human beings' dual relation to nature (insofar as humans are created out of dust) and to God (insofar as humans are created in the image of God). This dual kinship was exploited in an explanatory account of how humanity and the natural environment could form a harmonious relationship.

The essay entitled *Glimpses of Environmental Ethics from Indian Scriptures* by Dr. P.K. Bajpai and Dr. Nitish Dubey highlights the importance of *Srimadbhagawata Gita* and *Shukla Yajurveda*. The entire Vedic literature, throughout its texts, is marked with worship of Nature. The authors have tried to emphasize that Indian tradition has a rich heritage of environmental ideology. What is required is simply to interpret and understand the ideology, already given, in the present context of environmental protection. The present day attitudes of mankind, as West has led us in scientific endeavours and achievements, largely accords with Occidental ideology that all creation is for being used by humans and when such a concept becomes a leading one, the question of right to use becomes dominant over fading shadows of duties towards nature or in Vedic context the Creator and Creation relationship. The essays by Dr. Ranjay Pratap Singh, Dr. Sanjay Shukla and Dr. Shiv Bhanu Singh focus on the Vedic Hindu perspectives on human attitudes to nature. The Vedic perspectives are highlighted from the standpoint of the attainment of *moksha* (liberation) through the worship of nature in the form of *yajnas* and *mantras*. The cardinal virtues and values, i.e., love, compassion, non-violence, non-possession, contentment, etc., associated with Hindu Dharma can never permit us to exploit nature for our selfish ends. It is emphasized that Hinduism in its ritualistic and doctrinal aspects is a panacea for present environmental crisis.

The essays by Dr. Rajjan Kumar and Dr. Avinash Srivastava highlight the Jaina and the Buddhist perspectives respectively. The Jaina and the Buddhist perspectives spotlighted on the point of their ethos of incorporating measures for preserving natural resources into the overall effort to prevent the ethical downfall of human beings. Espousing the principles of *Ahimsa* and vegetarianism are demonstrated as examples of such environmentally sensitive ethical endeavours.

Using the windows of literatures, both traditional and contemporary, to view the relation between human beings and nature is the highlight of Section III on *Environment and Literature*. Glimpses of environmental wisdom reflected through the literary imagination must find a right place in the holistic approach to environmentalism. In this section, such glimpses are provided through the classical sources like *Ramayana* and through modern sources like *Kamayani* by Jai Shankar Prasad and *Asaadhya Veena* by Agyeya. In the essay entitled *Nature and the Poet: Observations on Tulsidasa and Valmiki*, Professor P.K. Pandey highlights the contributions of Valmiki and Tulsidasa and their portrayal of the reverential attitude that human beings had towards nature in our traditional Indian culture. It is held that the classical literary sensibility that visualized similarities between nature and human beings, both in internal and external aspects, is an expression of our genuine affinity with nature. This sense of affinity is, it is alleged, being seriously eroded by the strong impact of modern scientific culture.

Dr. Sanjay Kumar, in his essay entitled *Nature in Hindi Chhayawadi Poetry: An Instance of Jai Shankar Prasad's Kamayani*, highlights Kamayani as a unique poetical representation of oneness and unity between human beings and nature. Nature in Chhayawadi poetry is said to have been projected in different images: nature as divine; nature personified; and, nature as a moral teacher. This essay discusses Prasad's treatment of nature in Kamayani and emphasizes that Prasad not only conceives of nature as having an intrinsic worth, but also feels the need for man's communion with nature in order to get a glimpse of the Divine Spirit.

Dr. Sujata Chaturvedi in her essay entitled *Asaadhya Veena: A Poetic Symbol of Deep Ecological Self-realization* has tried to relate the theory of deep ecological self-realization to *Asaadhya Veena* written by the renowned Hindi poet Sacchidanand Hiranand Vatsyayana Agyeya. Agyeya in this text has exposed a deep sense of environmental harmony and importance of *atma-shodhan* (i.e., self-realization) and necessity of dilution of *aham* (i.e., ego) through this very meaningful philosophic rendering of literary thoughts.

The last and the fourth section of the volume concentrates on *Science, Society and Environment*. This section reflects upon multiple and diverse themes varying from energy resources, environmental standards, eco-feminism, sustainable development, pollution to the politics of environment. In the opening essay entitled *Energy Resources, Impact on Environment and Ethical Problems*, Dr. P.K. Rath states that given the culture of science and technology in the human world, environmental pollution to a certain degree is a fact of natural law. The supreme example that is cited is the human need for the generation of energy resources, a process that is bound to have a considerable amount of harmful impact upon the environment. Given this inevitability, recommendation is made for an imaginative use of waste management that would minimize environmental degradation. This recommendation harps on the imperative of our technological dealing with the environment in such a way that the overall natural process is manageably attuned to the so-called "biological clock".

In the recent times the need to evaluate the cost-benefit relationship for maintaining the regulatory requirements of environment has caught the attention of the international community particularly in the wake of globalization of trade and commerce. Mr. Vishnu Ratna in his essay discusses the role of management standards in improving the enviro.iment. These Environment Management Standards (EMS) are purely voluntary and do not carry the force of any legislative regulation; yet their adoption is becoming quite widespread in industry. Mr. Ratna emphasizes that an effective EMS can help an organization manage and improve the environmental aspects of its operations. Dr. Jyotsana Lal in her essay emphasizes upon water pollution as one of the most alarming public health problems in India in recent years. Based upon the empirical survey and research Jyotsana Lal states that water pollution menace can be overcome by community participation.

Dr. A.K. Sharma and P. Vigneswara Ilavarasan take up the question of dilemmas of sustainable development – the dilemmas of whether development measures at certain level of society are to be undertaken with unavoidable counter-developmental consequences at other levels of society, or avoid counter-developmental social consequences whatsoever at the cost of continuing social underdevelopment. It is pointed out that certain eco-sensitively sustainable developmental schemes are particularly subject to such a dilemma.

Dr. Vandana Asthana in her essay entitled *Development, Environmental Degradation and Marginalization of Women in South Asia* emphasizes upon the much talked about problem of eco-feminism, and the discussion on this hinges on a comparison of the exploitation of nature with domination of men over women. The eco-feminist position is crystallized by a striking allusion to the quality of caring and preserving inherent in the nature of a woman. This quality, it is argued, qualifies woman as the most desirable managers of natural resources in the administrative process of conservation and sustenance of the environment.

Dr. A. Saxena in his essay entitled *Environmentalism, Development and Human Security: An Obvious Connection* raises some important issues related to human security and sustainable development in the context of cost benefit framework. Taking India as a reference study for developing nations, Dr. Saxena pleads for alternative viable strategies for achieving sustainable development. It is held that solutions to the human security and development lie in the history of Indian Environmentalism.

Dr. Kanchan Saxena, Dr. Mukulika Hitkari, Dr. Deepshikha Banerji, Dr. Mrityunjaya Kumar and Mr. Purnendu Shekhar discuss various issues related to the socio-economic, moral and legal implications of environmentalism. A narrow, anthropocentric view of social justice is incompatible with environmental justice. Environmental justice, in the proper sense of the phrase, rests on the desirability of sustainable socio-economic development, which in turn demands the preservation of biodiversity in the ecosphere. Therefore, the ethical outlook is to be extended beyond the human-centred purview in a manner truly expressive of the sensitivity to environmental justice. This volume results from an attempt to understand the spirit of environmentalism within a holistic framework. However, the overriding motive of the volume is the articulation of the diffused contents of the sense of environmental wisdom. Such an articulation necessitates a multi-pronged reflective study of what the ideology of environmentalism essentially consists in.

I

Eco-centric Morality

1

Environmental Ethics:
A Historical-Intellectual Background

S.A. Shaida

The purpose of this paper is to provide a brief historical/philosophical background, which may enable us to understand and appreciate the rise of certain practical concerns and commitments in contemporary moral thinking. Till late fifties of the twentieth century these were not taken seriously by most of the moral philosophers. There are indeed some notable exceptions during the eighteenth and nineteenth centuries. However, the present survey is purported to situate applied ethics, of which environmental ethics is an important part, in the postmodern scenario where social-political concerns have the pride of place. To these we may add cultural and pragmatic approaches which enwiden the moral space and provide intellectual justification to deliberate upon human condition at large. The contemporary intellectual scene is sometimes in the risk of being understood, or dubbed, as anarchical on account of proclamations made, or epithets used, like "the end of ideology" during late fifties by men like Bell[1], "the end of the ethical" by Foucault[2] during sixties and "the end of history" by the liberal democracy and free market economy ideologue Fukuyama[3] during early nineties. This situation is somewhat confusing to those who are uninitiated into philosophy. Hence, this brief and modest attempt to clear the historical mist which surrounds the continuity of ethical thinking, keeping in mind the shifts which are specially in the realms of epistemology and social sciences.

II

During the Greek period we find certain themes and approaches that are still persistent in our age after having remained somewhat subdued in the medieval and in some stages of modern periods.

One idea, which is commonly shared by almost all important Greek thinkers, but most vocally by Aristotle, is that ethics is a practical science or study. The first group of thinkers who have placed ethics on the intellectual stage is of the Sophists. Protagoras' statement that "man is the measure of all things" is well known for its relativistic mode but it was also meant to be a principle to be inculcated among the citizens who were to be taught the art of dialectics and the nature of social-political rules which, if necessary, could be changed according to the existing norms. The main idea, which they sought to convey, was the social-cultural grounding of values and norms. They had to be meaningful in terms of the social order of the lifestyle they had chosen, of course, they could be influenced by rhetoric or sophistry, which has always been the mainstay of politicians and which the Sophists offered to teach at some price. As we will see later, a sort of such relativism, or subjectivism, which has often come to its rescue, was revived by Hume and Westermark and later by emotivists and most postmodernist thinkers. Plato offered a concept of good life that was supported by his concept of cardinal virtues, which could be imparted through his notion of knowledge as recollection. Presently, an anesthetized version of a good life finds expression in Foucault and others. For Plato, however, the knowledge of good life was possible as per the Delphic oracle communicated to Socrates through the well-known dictum 'know thyself'. This was generally interpreted as an identity between 'knowing good' and 'being good'. The interpretive bias in favour of theoretical knowledge as basis of good or virtue caused some problems in the understanding of Socrates intent. Much later, Gould in his celebrated work *Plato's Ethics* tried to establish the practical aspect of Socratic concept of knowledge as virtue. In Ryle's parlance, it was 'knowing how' rather than 'knowing that'.

This aspect of Greek ethics was further emphasized by Aristotle and was systematically distinguished from other forms of knowledge and practice. The highest form of knowledge was admitted to be *theoria*, philosophical/theoretical wisdom, which is pure speculative knowledge. Ordinary knowledge, both deductive and inductive, was called *episteme*, which he contrasted with *phronesis*, which means practical reasoning or wisdom. *Techne* for him is practical/fabricating skill. Keeping in view the importance of *phronesis* in his ethical thought, we can appreciate the role of practical syllogism in the process of moral reasoning. The opening lines of Aristotle's *Nichomachean Ethics* sets the practical tone of his ethics. As he says, every action of a man has an end or aim and therefore the aim or end of action is good for man.

Concurring with Plato's tripartite division of soul as rational, active and appetitive and accepting the rational (with the above stated distinctions) as the highest, moral action falls within the territory of *phronesis.* In this context, he talks of virtue as a habit of choice according to relevant deliberation. One important guide in making right choices is the principle of means, which avoids both the excess and defect. Acting on such rules and principles, one can achieve *eudemonia* or happiness that, for Aristotle, is the *summun* human. It is a sort of ethico-aesthetic life dedicated to the commitment of achieving excellence (i.e., virtue) in whatever a man decides to do according to his rational choices. The above brief account of Greek ethics presupposes that it was rooted in the prevalent Athenian ethos, which made certain virtues, especially justice, necessary to maintain social order of that age. In the brief canvas of this paper it is not possible to go into details of this interesting theory nor into some very objectionable and crucial aspects of Greek culture such as strong pro-slavery and anti-feminist stance which is subtly embedded in Greek ethics and epistemology.

The medieval period primarily advocated theo-centric ethics based on Christian theological expositions mainly by St. Thomas Aquinas and St. Augustine. Their theoretical interest did however permit some discussions on war and some other relevant issues in that age.

During the modern period, philosophy developed on two different epistemological foundations in Great Britain and on the Continent. The British philosophy was rooted in empiricism, which justified the search for ethical foundation in the understanding and analysis of human nature as well as in the prevalent social-political milieu. The echo of the Continental and especially Cartesian attempt to wean philosophy away from ecclesiastic domination can also be heard in the empiricist discourse of Hobbes, Bentham, Mill, Hume and others. A natural fall-out was a distinct approach in ethics that was dominated by hedonism and utilitarianism. Besides human psychology that by and large established the supremacy of desire for pleasure, there was a strong underpinning of political situation as existing at that time. Hobbes' plea for unreserved obedience to the *Leviathan* who could ensure security and peace to the brutish and selfish mass and Bentham's and Mill's attempts to bring about legislative reforms based on utilitarian principle were actually the search for social-psychological foundation of ethics. They all, including

Locke, were operating within social-contractual model of the relationship between individual on the one hand and society and state on the other. Hobbes' *Leviathan,* Locke's essay *On Government* or Mill's essay *On Liberty* belong to the same genre of ethico-political writings. Psychologically, they drew support, though sometime for different reasons, from their understanding of human nature, which brought out the significance of survival instinct or fear (Hobbes), desire for pleasure (Bentham and Mill) and feelings and sentiments (Hume). Later, even Butler *(Sermons)* developed his intuitionistic theory in the light of his vision of human nature. Sidgwick *(Methods of Ethics)* largely depended on human psychology for his account of utilitarianism in terms of rational egoism, rational benevolence and justice. For utilitarians like Bentham and Mill, in the context of possible conflicts between egoism and altruism, the society and the government do and should intervene with the help of certain sanctions with varying force and pressure. This was theoretically built in Bentham's *Principle of Morals and Legislation*, which Mill endorses in his *Utilitarianism*. The social-political interest of these thinkers brought practical concern of the Greeks once again to the fore. Some of them wrote on the state of prisons (which was also a matter of concern for Dickens), punishment, the condition of women or suicide.

On the Continent, the Cartesian philosophy set the rules and principles of the rationalist discourse, which was followed by Spinoza and Leibnitz. Descartes was chiefly interested in establishing the indubitable rational foundation for all forms of philosophical reflection though he remained largely confined to epistemological ad metaphysical issues. He is credited with breaking away from the ecclesiastical control and establishing the authority of reason, which alone can offer self-evident *a priori* truths. Human self or *cogito* was installed with full pristine glory to the throne of philosophical reasoning. The ego-centricity embedded in his philosophy can be interpreted as the first revolution anticipating, and also being similar to, Kant's *Copernican Revolution*. Descartes' attempt at founding all philosophy on clear and distinct ideas paved way for *a priori* reasoning in moral philosophy as well which was taken up not only by Spinoza on the Continent but also by men like Wallaston, Wolf and Clarke in England. Hence, the presences of some objective, an immutable, absolute *a priori* ethical principle was felt inevitable and was derived often in collaboration with metaphysics. It was, however, Kant in whom ethics found a very convincing champion of rationalist ethics. In his *Groundwork of the Metaphysics of Morals* and *Critique of Practical Reason*, some epoch-making concepts were formulated which dominated ethical thought for a very long time. His concepts of goodwill, the autonomy of the will, the universality of moral law and the categorical imperative jointly provided a superarching model or paradigm for all ethical theorizing and a rational foundation for all morality. Kant derived all these principles *a priori* from what he termed *practical reason*, which was not at all practical in Aristotelian sense. It was only a reason in search of some presuppositions which could neither he supplied nor justified by *pure reason* which could offer only *analytic* propositions that are empty of content. Thus, Kant's *practical reason* paved way for procuring foundational judgements for metaphysics, religion, science, ethics and aesthetics. The most important theme underlying Kant's ethical theory was his view of freedom or autonomy of the self-legislative will which must always be jealously guarded against heteronomy, i.e., any external, non-rational intervention that is not sanctioned by the rational will. Thus, all psychological/empirical motivation was out of bounds since the rational will has no control over them. The only rationally justifiable motive for doing any moral act is duty itself. Such a formalistic theory had no concrete guidance for moral actions; it only supplied a principle determining *how* a moral action ought to be performed. The only specific injunction available is that one should treat humanity, including oneself, always as an end and never simply as a means. Nevertheless, there was a general euphoria at the triumph of reason which was celebrated in Kant's own assessment of the establishment of rationality in his essay *What is Enlightenment?*[4] For Kant, it was an 'ongoing process', a 'task and an obligation' which the rational being *qua* rational is fitted as well as obligated to carry forward as a mark of progress. The age of enlightenment appeared to be the joint vindication of rationalism in ethics and religion on the one hand and of science as developed by scientists such as Copernicus, Kepler, Bruno, Galelio and Newton on the other. The rationalism of Kant found its logical culmination in Hegelian philosophy where ethics was placed on the metaphysical pinnacle of spiritual self-realization. The absolutist turn of rationalism into idealism reigned supreme on the Continent for more than a century. During its last phase it received a short lease of life in Italy and England. At Cambridge, Bradley, Green and Bosanquet ruled through McTaggart's influence till the end of the nineteenth century.

During the idealistic hegemony the practical import of ethics was by and large marginalized. Hegelianism and neo-Hegelianism remained as speculative as metaphysics. Hegel's historicist account of the development of human self was a gradual progress of man only at the level of pure thought whose dialectical development moved towards more and more totalization, which culminated in the metaphysical supremacy of the state, which deprived the individual of all freedom. The self-legislating will was buried under the burden of the oppressive power of the state.

The advent of the twentieth century saw a radical shift in ethical thought. Even before the rise of logical positivism/empiricism and the prospective impact of early Wittgenstein, Moore's *Principia Ethica* (1903) introduced the linguistic/conceptual analysis in explicating the nature of moral concepts. His formulation of the naturalistic fallacy remained a dominant theme in ethical discussions for almost half a century. But it would be a mistake to consider Moore's ethics as purely analytic or meta-ethical. His ideal utilitarian statement of the ideal makes it a significant addition to the body of normative ethics. Some other thinkers like Prichard and Ross carried forward the analytic programme. During the forties of the last century meta-ethics proper prevailed over the Anglophonic ethical thought under the dual influence of the *Tractatus-Logico-Philosophicus* (including logical positivism) and *Philosophical Investigations*. Ayer's *Language, Truth and Logic* (1938) exemplifies the former while Stevenson's *Ethics and Language* (1944), Toulmin's *Place of Reason in Ethics* (1950), Hare's *Language of Morals* (1952), Nowell-Smith's *Ethics* (1954) and some other works belong to the latter category. Most of these works played upon Wittgenstein's concept of language games but failed to exploit the deeper significance of his views concerning language as a 'form of life'. They did talk of close relation between ethics and conduct but the practical concern was given a linguistic turn by suggesting that the function of ethics (i.e., language of ethics) is to persuade, convert, command or to prescribe. Ethics, essentially, remained purely meta-ethical, which was not allowed to "contain any normative judgement" (Ayer) because it was only a "logical study of moral language" (Hare).

III

During mid-twentieth century and after, various philosophical and intellectual tendencies came to fore which may explain a sort of rupture or discontinuity in philosophical tradition and particularly in the ethical discourse, which is our main concern. As it is a stupendous task to go, even with brevity, into all the forces that had metamorphosized contemporary intellectual climes and concerns, we can only point out some of those, which are responsible for ushering in the preponderance/revival of the practical approach in ethics.

Among a number of thinkers, including some meta-ethicists themselves, there was a growing realization that the clarificatory task of the linguistic philosophy was quite overdrawn and complete banishment of normative issues from the realm of ethical thought was getting out of place in the light of fast developing contemporary changes. Specific moral issues could also be analyzed and discussed without assuming the role of moral preachers. Philosophers, being the denizens of social-political world, had also an obligation to discuss and understand the reasons/causes and impact of various events and phenomena occurring in our societies. The attempt at self-reflection saw the publication of *Clarity is Not Enough* (ed. H.D. Lewis). Hare also shows a bent towards taking some practical issues in *Freedom and Reason* (1962). Russell's book *Commonsense and Nuclear Warface* (1959) and his other writings and the intelligentsia took discourses on disarmament and sexual morality seriously though it had little impact on professional philosophers. Hannah Arendt's *Human Condition* (1958) was also an attempt to draw our attention towards trends and developments in society-affecting man's subjectivity and freedom. At the same time, as C. Wright Mills[5] pointed out, historical evidence was manipulated to manufacture "a trans-historical straight-jacket" and social sciences were prescribed the construction of "a systematic theory of the nature of man and society". For many, such an attempt could only lead to 'the broken image' or 'one-dimensionality' (Marcuse) of man. The same fear was expressed by the famous British historian Sir Lewis Namier in his works *England in the Age of American Revolution* (1930) and *Personalities and Power* (1955). Hence, attempts started to situate human self in its contextualized socio-moral and psychological particularity, de-linking human action from any superarchical rational model like Kant's. This anticipates the rise of what we today call *applied ethics*. Besides, a number of thinkers like Rawls[6], Bernard Williams or Michael Walzer also had the same complaint against Kant or other foundationalist approaches where, universal, rational and objective principles offered the transcendental peg on which every ethical action has to be hanged. Williams argues for the way of thinking in ethics, which breaks with Kantian, or other formalist approaches, giving proper place to contextual factors and culture-specific values and interests, which have great significance for moral agents.[7] Walzer also pleads for respecting varying values and viewpoints rooted in cultural 'forms of life', which coexist in any genuine democracy.[8] Delinking themselves from Kantian *a priorism* and foundationalist stance, some of them like MacIntyre[9] and Williams incline towards Aristotelian preference for practical virtues in which lies the concept of good life manifested in civic and private perfection. The disjunction between private and public morality appears redundant in practical morality. Hence, "the end of the ethical" only means the end of the search for universal, objective, absolute and trans-cultural values. The end of ethical should not be considered moral nihilism. It only asserts that moral thinkers should now attend to particular moral issues, which arise in human society and politics with their multidimensional contextually. The coalescence of social, political, religious or moral is often evident in problems like suicide or female infanticide.

IV

What we have said above indicates the due necessity of some self-reflection. Various thinkers have drawn people's attention towards specific moral problems, which arise in private and public domains. Modern society is facing terrible emotional and other psychological strains besides numerous socio-economic and political challenges. Above all, the ecological degradation and disequilibrium are haunting us with their prospect of irreparable damages to nature as a whole, including the entire race of *Homo sapiens.* This situation claimed serious concern and understanding, which revived the interest in practical-moral reasoning towards our existential problems. The shift from the ethical to the moral can be interpreted as trans-evaluation of values. Initially, it was variously described as practical, contextual or situational ethics. Now, it is termed *applied ethics*, which includes business ethics, legal ethics, medical ethics, bio-ethics or environmental ethics. Some of the specific issues, which have come in close scrutiny and are being seriously discussed, are suicide, euthanasia, abortion/infanticide, pro-natal sex determination, female foeticide, punishment, slavery, anti-feminism, or genetic engineering. Within environmental ethics, some important problems are environmental degradation, deforestation, air, water and sound pollution, or destruction of ecological balance. Some problems like overpopulation and nuclear arms race and war are multidimensional and are often covered by various branches. One of the undercurrents in most of these discussions is the utilitarian approach. This is evident in the works of Sen and Williams[10] or Smart and Williams[11], which bring out different aspects and problems in applied ethics. Sen's work on poverty and famine[12] has great relevance to countries in the South or the Third World. The widespread interest of contemporary thinkers in such problems led to the publication (1971 onwards) of journals like *Philosophy and Public Affairs, Journal of Applied Philosophy* and *Environmental Ethics.*

In order to provide proper perspective and adequate emphasis, men like Arne Naess[13] (who popularized the idea of 'deep ecology') and

Johnson[14] focus upon the general approach towards environment. They try to highlight some of the disastrous consequences of what we are doing to the environment without realizing that what we are doing to nature as a whole is actually, what we are doing to ourselves. As Johnson points out, environmental degradation is a time bomb, which can explode any time if the present suicidal attitude of human beings continues unabated. What is needed is the all-significant paradigm shift from anthropocentricity to eco-centricity. According to Johnson's holistic approach, the ecosystem as a whole is to be regarded as a great organic whole comprising all organic and non-organic systems. From cosmic point of view, each and every system—animals, plants or other things—has its own place and value. Destruction or degeneration of one/any, artificially brought about, starts a chain reaction, which disturbs the equilibrium of the ecosystem as a whole. It is not only things on the earth, but the total atmosphere above and around it, which form the ecosystem. The harmful consequences of the depletion of ozone layer, leakages of poisonous gases or atomic radiation, spread of excessive carbon dioxide through automobiles and other gases discharged by industrial units are well known. We have not forgotten happenings like Hiroshima and Nagasaki, Bhopal gas tragedy or Chernobyl leakage of atomic radiation. Singer and Regan[15], Passmore[16] and various other thinkers[17] have discussed these and other related issues at length.

In India, awareness of these problems is at a very low level and governmental measures to tackle these are extremely inadequate. Sometimes, they appear to lack proper sensitivity as well as political will (e.g., in case of starvation deaths in Orissa). Some measures adopted recently like the deployment of CNG buses in Delhi, banning public smoking or burning of crackers beyond certain hours, removal of polluting industries from Delhi are not because the government was serious about them but they were directives issued by the Supreme Court either *suo moto* or in response to public interest litigation. Inclusion of environmental problems in university curricula is also one such measure, and not a brainchild of Ministry of Human Resource Development. There are sincere and committed individuals like Sunder Lal Bahuguna (of Chipko Movement) or Medha Patkar and Arundhati Roy (of Narmada Bachao Andolan) whose attempts are often understood by the government as obstructive in the way of development. It is, nevertheless, such people and various NGOs who can effectively achieve some success.

The role of media in this attempt is very important. And, I mean the print media, not the electronic one, which can be broadly categorized into *Sarkar Darshan* and *Lakshmi Darshan*. Newspapers can, to a large extent, disseminate and guide the social awareness and enlighten the masses regarding our obligations towards the ecology with all its constituents—animate or unanimated. If not for the sake of cosmic unity, at least, from the anthropocentric point of view, for the sake of human survival and posterity, its importance can be brought home. If we have any concern and love for our children and grandchildren and for their children and grandchildren, the writing on the wall must be properly seen and communicated. We have a large number of committed journalists and media persons who can take up such issues. Here I see greater role of newspapers in Hindi and other regional languages that reach the majority of the Indian masses.

Finally, I consider the possibility of our young men and women, mostly students and their organizations, which can perform this function most effectively provided they refuse to become pawns in the hands of corrupt and self-interested politicians. Many of us would recall the students' protest in Paris, a few decades ago, which was led, by a philosopher and a student leader Sartre and Tariq Ali. In India, recently, I saw how students in the North-East region completely and effectively banned felling of the trees and now no one can dare cut the trees anymore.

To conclude, what is to be seen is that the attitude of conquering nature or achieving victory over nature is to be given up as a debunked ideology of the technologies of Western mind. We need a paradigm shift from 'victor-vanquished' model to the Buberian 'I-Thou' paradigm. What we need is a sort of ecosophy, which, in the words of Naess, "can provide a single motivating force for all the activities and movements aimed at saving the planet from human exploitation and domination".

Notes

1. Bell, Daniel, *End of Ideology*, New York, 1960.
2. Foucault, M., *The Order of Things*, London, 1970; *History of Sexuality*, Vol. I, London, 1979.
3. Fukuyama, F., 'The End of History', *The National Interest*, Washington D.C., Summer, 1989; *The End of History and the Last Man*, London, Hamish Hamilton, 1992.
4. Kant, I., 'What is Enlightenment?' in L.W. Beck (ed.), *Kant on History*, Bobs-Merrill, 1963.
5. Wright Mills, C., *The Sociological Imagination*, New York, 1959.

6. Rawls, J., *A Theory of Justice*, Cambridge, Mass, 1971; 'Justice as Fairness', *Philosophical Review*, Vol. 72, 1958.

7. Williams, Bernard, *Ethics and Limits of Philosophy*.

8. Walzer, Michael, *Spheres of Justice*, New York, 1983.

9. MacIntyre, A., *After Virtue*, Duckworth, 1981.

10. Sen and Williams (eds.), *Utilitarianism and Beyond*, Cambridge, 1982.

11. Smart and Williams (eds.), *Utilitarianism: For and Against*, Cambridge, 1973.

12. Sen, A., *Poverty and Famines*, OUP, 1981.

13. Naess, A., 'Deep Ecology and Ultimate Premises', *Society and Nature*, Vol. I, No. 2, 1992; *Deep Ecology in Good Conceptual Health*, The Trumpeter, Fall, 1986.

14. Johnson, L.E., *A Morally Deep World*.

15. Regan, T. and Singer P. (eds.), *Animal Rights and Human Obligations*, Englewood, 1976.

16. Passmore, J., *Man's Responsibility for Nature*, London, 1974.

17. There are some other good works on these problems like:

    (a) Singer, Peter, *Practical Ethics*, 1979.

    (b) Singer, Peter (ed.), *Applied Ethics*, OUP, 1986.

    (c) Racjel, James (ed.) *Moral Problems*, New York, 1971.

    (d) Nagel, T., *Moral Questions*, Cambridge, 1979.

    (e) Goldwitch, Goldwitch & Harris, J. (eds.), *Animals, Man and Morals*, 1972.

    (f) Parfit, D., *Reasons and Persons*, 1984.

    (g) Walzer, Michael *Just and Unjust Wars*, New York, 1977.

    (h) Shrader, Frechelle, *Environmental Ethics*, 1981.

    (i) McCloskey, H.J., *Ecological Ethics and Politics*, 1983.

    (j) Elliot and Gare, *Environmental Philosophy*, 1983.

    (k) Wassertorm, R. (ed.), *Today's Moral Problems*, 1979.

    (l) Narveson, J., *Moral Issues*, 1983.

    (m) Worster, *Nature's Economy: A History of Ecological Ideas*, Cambridge University Press, 1977.

    (n) Miri, Sujata, *Ethics and Environment*, Spectrum Publication, Delhi, 2001.

    (o) Naess, A., *The Shallow and The Deep, Long Range Ecology Movement;* and *Bio-spherical Egalitarianism*.

2

On the Adequacy of the Human-centric Model of the Field of Moral Relations

Rajendra Prasad

The word 'field' in the title of this essay and in the discussion that follows is used in the strictly logical sense in which the field of a relation is the sum of its domain and converse domain. There is a traditional, prevalent, or common sense, model or the field of moral relations according to which, ordinarily, the field of all moral relations consists of only human beings possessing a certain amount of rational and discriminative maturity. This means that only a human being can, conceptually speaking, be the domain or converse domain of any moral relation. In the last two or three decades, moral philosophers have started taking a lot of interest in areas like environmental ethics, animal ethics, etc., as some of them feel that the traditional model of the field of moral relations, or to put it briefly, of the moral field, is suitable only for talking about, or conceptualizing, moral relations existing between human beings, or between human things, and not between human and non-humans, or between one non-human and another. Therefore, it is unsuitable for talking about, or exploring, any issue pertaining to such areas as environmental ethics, animal ethics, etc., in which some reference to something non-human is unavoidable. I intend to show in what follows that there is nothing wrong with the traditional model; that is, nothing wrong with holding that normally, standardly, or ordinarily, as per the way we understand the concept of a moral relation, only a human being can be its domain or converse domain. This model has built into it some sort of flexibility, or openness, taking advantage of which we can extend, if we need to, our notion of a moral relation, ordinarily existing between humans, to talk about any ethical concern involving a moral relation, or something like a moral relation, to a non-human thing, event, or phenomenon, etc. Therefore, I would conclude that the traditional model of the moral field is comprehensive enough to take care even of such areas of ethics in talking about which we have necessarily to deal with some non-human thing.

I shall first briefly unfold the natural evolution of the traditional model involved in the very logic of the concept of a moral relation.

Traditional Model as Human-centric

As already said, the field of a relation is the aggregate of its domain and converse domain. For example, in the sentence "x is greater than y" all those constants, any one of which can meaningfully replace x would form the domain and all those which can meaningfully replace y the converse domain of the relation 'greater than'. Similarly, in the case of the moral relation 'fair to' in the sentence "x is fair to y" anyone who can be fair to someone would belong to its domain and anyone to whom someone can be fair to its converse domain. The field of a relation being the totality or sum of its domain and converse domain, the field of the moral relation 'fair to' would thus be the totality of all those who can be fair to anyone and of all those to whom anyone can be fair.

At least a good number of moral relations are such that one can have them to oneself as well as to someone else. One can be fair (or unfair) to oneself as well as to someone else and 'being fair (or unfair) to' denotes, at least in some uses of it, a moral relation.

A moral relation can operate on two levels, which may be called non-participatory and participatory. It works on the non-participatory or assessive level, for example, when one judges or evaluates an action to be right or wrong, commendable or condemnable, morally relevant or irrelevant to the context in which it has been done, etc. He can evaluate in the moral sense, or from the moral point of view, not only actions, but also intentions, persons, groups of persons, institutions, etc. In doing this he enters into the relation of evaluating or assessing the moral value of the object judged. To evaluate the status or role of anything is to enter into a relationship with it, of course, on the mental level. The relationship is moral, whether one evaluates the moral worth of the thing in positive or negative terms, since the evaluation done may itself be morally right or wrong, justified or unjustified. Judging it right to give to a man of a certain caste a punishment harsher than the one given to a man of another caste when the crime committed by both of them is of the same moral status, is not only logically wrong but also morally wrong. To judge the moral worth of a thing is to adopt an attitude towards it, which may be an attitude of approval or disapproval, commendation or condemnation, praise or blame, etc. And, the adoption of this attitude can itself be morally defensible or indefensible. Therefore, to call the relation of judging or evaluating a moral, i.e., morally judge able, or criticize able, relation is quite in order. To put it in another way, judging the moral worth of a thing is to do something intentionally. Rather, it is something that cannot be done accidentally, inadvertently, or in a huff because it requires one to examine whether or not the thing being evaluated satisfies the relevant criterion or criteria (whatever they may be) of rightness, obligatoriness, or goodness, etc. Such an action has to involve self-consciousness, and therefore cannot be accidental or unintentional.

A moral relationship with the object morally evaluated can be held only by one who possesses the ability of judging, who has some idea of the criterion, or criteria, ground or grounds, i.e., some idea of the reasons, for judging its moral status or worth in positive or negative forms. That is, he has to have a certain level of rationality, or rational maturity. This means that only a human being can be such an evaluator. No plant, or non-human animal can be credited to have the rational maturity or ability to judge the moral worth of anything. It is obvious, then, that the domain of any (moral) relation of morally judging must consist only of human beings possessing a certain kind of rationality, or reasoning ability.

A moral judge may not do anything besides judging, or judging and publicly expressing his judgement. But he may also become a moral activist, that is, proceed to taking the necessary steps called forth by his assessment or understanding. Many Indians considered it morally wrong for an elected legislator to continue enjoying the status and consequential benefits of a legislator even when he has no care or concern for the welfare of the people who have elected him. But Jai Prakash Narayan became a moral activist by campaigning for giving to an electorate the power to de-elect, before the expiry of the period he has been elected for, an elected representative of theirs if and when he has failed to perform the duties he, as a legislator, is required to perform. When one starts taking action in the light of his judgement, or holds a substantive moral relation with anyone, his moral relation with the object or objects concerned becomes participatory in the sense that he participates in a moral process or transaction. Jai Prakash Narayan's relation with the then legislators, whom he considered non-functional or mal-functional, became participatory when he started his movement, which, though taken to be largely political, was also a moral movement.

Even in ordinary transactions one enters into a participatory moral relationship when he does something, which can be morally judged. For example, when a man distributes his parental property among his brothers and sisters strictly in accordance with their bonafide claims, he is fair to his brothers and sisters. 'Being fair to' is a participatory moral relation. So would be its opposite, i.e., 'being unfair to' . Other examples of such a relation would be 'being cruel to', 'being kind to', 'being faithful (or unfaithful) to', 'being polite (or rude) to', 'being respectful (or disrespectful) to', 'being sincere (or insincere) to', 'being considerate (or inconsiderate) to', 'being benevolent (or malevolent) towards', 'being caring (or calous) towards', 'being of a forgiving (or unforgiving) nature towards', 'being friendly (or unfriendly) with', 'being altruistic (or selfish) in', 'being sensitive (or insensitive) to someone's suffering', 'being grateful (or ungrateful) to'. etc.

It is also possible that one participates in a moral relation and later on reflects on it and judges it to have been rightly, or wrongly, executed, as we do in self-criticism or self-appraisal. This means that the same person can hold both, a participatory and a non-participatory, relation.

When we consider anyone of these relations, or any of their ilk, we notice that to hold a moral relation to anything requires a certain kind of mental maturity which is not expected of any non-human thing. Take for example, the moral relation of being faithful to. When we say of a husband that he is faithful to his wife, we assume that he understands the status of the conjugal relation among human relations, the obligations which accrue to him after entering into this sort of relationship with a woman, the social approbation which goes with one's being faithful, and social dis-approbation which goes with being unfaithful, to his wife, the contribution which faithfulness makes to one's leading a happy conjugal life or to having a happy home etc. This kind of understanding we cannot expect a plant or a non-human animal to have. Not even of a dog though we take it to be an unquestionable truth that a dog is faithful to his master or mistress. What I want to emphasize is that mental caliber which being faithful to requires is not exhibited by a non-human animal. We notice some similarities between the behaviour of a domestic dog and that of a human being, say, an and servant, and say that the dog is faithful, or even that the dog is more faithful than the servant, to its master. The difference is that the behaviour of a dog is instinctive, or conditioned, while that of a human being thought-out or reasoned. A faithful behaviour of a husband is not only inspired by his love for his wife and his own awareness of the obligations of the marriage-bond but is also a response to her behaviour towards him. It is not only responsive but also discriminative. It is discriminative in the sense that a faithful husband discriminates between a behaviour of his wife which is a proper response to his faithfulness and one which is a sign of betrayal. A faithful husband would not mind his wife's letting her brother kiss her. But, a faithful dog of the lady would pounce upon her brother if it has not till then been made by her friendly with him. What I want to underline is that not only the judgemental relation but also other moral relations, can be held only by human beings, and not by any plant, any non-human animal, or any inorganic object, like a rock, a river, or desert, etc. This means that the domain of a moral relation can consist only of human beings. Its boundary is the boundary of the human. In this sense, the domain is well-bounded.

To say that only a human being can occupy a place in the domain of a moral relation is not to say that all human beings can do that. To hold a moral relation, as has been said, one has to have some mental, or rather rational, maturity, or ability. For example, the moral relation of being responsible for cannot be held by an infant, an idiot, a subnormal, or a mad man. But the requirement of rationality in a person for holding a moral relation is flexible, and not too rigid. It is flexible in the sense that it can be held by persons of different levels of rationality. For example, at the bottom, we can have a teenager resenting his mother's not caring for him because of her spending a lot of her time in managing the affairs of the ladies' club she has founded, and at the top end the highly qualified, well-studied, individual equipped with the ability to examine pros and cons of every decision he takes. The latter would be someone the like of whom Butler has in mind when he speaks of the cultured, decent, Englishman, as an example of a moral agent whose self-interest overrides an impulse of his, and his conscience overrides his self-interest, when there is a conflict between an impulse and a self-interest, or between a self-interest and conscience. In between the two ends we can have persons possessing differing degrees or levels of rational maturity.

Let us now turn to the converse domain of moral relations. That one human being can have a moral relation with another is obvious. And, normally our talk of moral relations is a talk of a kind of social relations, which can exist between one man with another, or between one group of men with another. I can, for some moral reasons, resent, question, condemn, avenge, appreciate, admire, express gratitude for or forgive, etc., a certain action of my neighbour, and to a lot more of several moral things in my dealings with him. The assumption here is that I can have a full-fledged moral relation only with such a being with whom I can communicate and he can communicate with me, who understands the meaning of what I do and I understand what he does, i.e., I can interact with him, argue with him, try to convince him of my stand, or be convinced by his argument, etc. Entering into a moral relationship is thus to participate in a social transaction.

Our moral relationship with children may seem to pose a problem for what has been said above because we cannot have with the kind of communication we can have with an adult. But still we have duties towards them. One the other hand, children have rights on us, but no duties towards us. They cannot have any duty because they do not possess the maturity which being dutiful requires. A mother cannot deny that it is her duty to feed well her six months' old daughter, but she cannot say that the child has the duty not to cry and disturb her when she is engrossed in reading an interesting novel. She cannot also deny that child has her rights on her, for example, the right to be fed well, to be kept clean, etc., though she has no duties towards her (or anybody else). But this case does not pose any problem for the traditional model because one's having a right on someone does not always imply him owing a duty to the latter, nor does one's having a duty towards someone always imply his having a right on the latter. The mother has no right to be fulfilled by her child daughter, though she has some duties towards her, and the daughter has right on her mother but no duties towards her. The moral relation of being duties towards the child is possible because of the assumption, based on the empirical fact, that a child has the potentiality to grow into a normal individual, a full-fledged member of the moral community with whom one can have social interaction. This is not the case with any non-human thing. We cannot assume that a plant, a hill, or a cat, would grow into an individual, a moral agent.

Sometimes we speak in a language, which gives the impression that we can have a moral relation also with an animal. A housewife is angry at her pet cat for having drunk her bedtime milk, and so she is also with her servant who has done a similar thing some other night. We may say that both are guilty of a similar offence and the housewife's anger in both the cases is moral anger, i.e., anger felt for a moral wrong done one night by the cat and another night by the servant.

When we examine the two culprits and their faults, we find that the anger on the cat's having drunk the milk cannot be called moral anger, nor can the cat be called guilty of a moral wrongdoing. The cat has no idea of what is right or wrong; she does not have a reason for drinking the milk, while the servant may have one, say, the reason that the housewife does not give him enough to fill his stomach. Moral anger is felt on one's intentionally doing something, which he could have avoided, and for doing which he does not have a justifying reason. If there is a justifying reason for doing it, then it becomes morally right and therefore the anger felt for having done it would not be a moral anger. The cat drinks because it is her nature, instinctive propensity, to drink milk if it is reachable to her; she does not deliberate over, or think of, the desirability or undesirability of drinking her mistress' milk. The housewife cannot convince her of having done something wrong because no communion with her is possible. The housewife can surely be angry with the cat because she has suffered a loss, but her anger would be simple annoyance and not moral anger. Reasons like those mentioned above have given rise to the traditional or commonsense view that moral relations can hold only between one human being and another and not between a human being and a non-human animal, plant, or an inorganic thing like a hill etc. Ethics has been, thus, traditionally defined as the study of what a human being living in a society, ought to do or ought to be, which in effect means a study of human relations from the moral point of view which may be deontological or teleological. Thus, the traditional, or prevalent, model of a moral relation presents it as a relation both the domain and converse domain, i.e., the field, of which consists of human beings possessing, or having the potentiality to possess, a certain kind of rational, or deliberative, decision-making, ability.

Human-centric Model as Accommodative of Extending Moral Relations to Non-humans

The human-centric model does not necessitate or entail that a human being cannot have any moral relation to their environment, to the plants, animals, rivers, hills, etc., which occupy a large part of his world and which are useful to him in so many ways. As per the traditional model, though a moral relation can straightforwardly exist only between one human and another, we can still say that one ought to take good care of his environment or surroundings. We say one ought to protect his forests (because they cause rainfall), keep the rivers unpolluted (because they give him water needed by his plants, animals, crops, etc.) and ought to be considerate towards his animals (because he benefits from them in so many ways). But, as it is clear from the because clause in braces in each of the above examples, his obligation to non-humans arises from the fact that the non-humans serve some of his interests or needs, or are usable as means or sources of deriving some benefits.

During the last three or four decades of the twentieth century moral philosophers have started paying a lot of attention to such areas as environmental ethics, animals ethics, bioethics, etc., which are concerned with man's ethical attitudes towards some non-human things or aspects of the world he is in, or of his surroundings. In the terminology of this essay, these are ethical concerns in which the converse domain of a moral relation is something non-human. In the traditional, human-centric, approach to ethics, as we have seen, the entire field of moral relations is held to be human. Therefore, some modern ethicists think that we need a new ethics, or a new approach to ethics, to enable us talk about the issues pertaining to these new areas of ethics in which the converse domain of a moral relation is something non-human. But these areas do not pose any serious threat to the human-centric, traditional, model. The latter is flexible or accommodative, enough to enable us to express our sensitivity to the ethical importance of rivers, plants, animals, etc., while remaining within its framework, we can hold and talk about a moral relation to a non-human thing because the non-human converse domain of such a relation has always a link, as will be shown below, with some human interest, i.e., something human.

A non-human thing, no matter howsoever greatly we value it, is always of only instrumental value, or speaking more broadly, of value only because it satisfies some human interest. We say that a farmer ought to keep his ox clean, say, because then it would not transmit any infection to him; that a bus driver ought to see that his bus does not emit gas while playing, because if it does it would pollute the local air and cause breathing trouble to those on the road who happen to inhale that air; that a hill dweller ought not to denude a hill of its forestation because forestation helps rainfall which men need for cultivation, etc. Thus, in such cases too, the converse domain of the moral relation of 'ought to', or of 'being obligated to', apparently between a human agent (a farmer, a bus driver, a hill dweller, etc.) and a non-human thing (an ox, a bus, forestation, etc.) is ultimately between a human agent and a human interest-satisfying thing, i.e., something human. This shows that the human-centric model can take care of all such cases.

In the case of a virtue like kindness, it may seem that it can be said that we ought to be kind to animals without having in mind any interest of ours to be satisfied by being kind to them. It may not be always so because kindness is generally required towards those animals which serve some human interest and not towards those who pose a danger to human existence, or are prone to cause some human suffering. But, even if we admit that we ought to be kind to an animals, no matter whether or not being kind to it serves any human interest, this admission does not disprove the adequacy of the human-centric model.

Kindness is a response which is in order only when it is felt towards a being which is experiencing, or subjected to, some suffering, more specifically, to one who is suffering some pain, or has the sensitivity to suffer pain. Therefore, there is no point in being kind to rocks or rivers. But, an animal has the sensitivity to suffer pain, and in this respect animals and human beings are similar. It is this similarity, which makes kindness to animals a virtue, as is kindness to children or sick men. In requiring one to be kind to animals we assume animals to be similar to human beings. Therefore, kindness to them is kindness to human-like things, and consequently covered by the human-centric model. In fact, anything which one considers to be human-like, say, a plant, a piece of furniture, a book, a river, etc., can be an object of his kindness. Shakuntala, in Kalidasa's *Abhijnana Shakuntalam*, feels very kind not only to the pregnant deer which is to deliver its baby in a few days, but also to the plant and climbers she everyday waters because she thinks that all of them can suffer pain, and in this respect resemble humans.

Sometimes we do argue for preserving a species of animals which are dangerous, or one of plants which are poisonous. Even, in such cases, there is a concealed, sometimes openly declared, reference to some human interest. Generally, we want to preserve them because we can study them only if they are preserved and are readily available. Here, the human interest involved in the interest to satisfy our curiosity to known what kind of things they are and whether or not any use can be made of them. When we argue for preserving nature on the ground that we enjoy it, obviously it is a human-centric argument. We can conclude, therefore, that in any ethical enterprise, whether it is concerned with human or non-human subjects, there is no escape from accepting the primacy of the human, or from adopting the human-centric point of view.

3

Environmental Wisdom

Bijoy H. Boruah

In his Lloyd Robert Lecture delivered at the Royal Society of Medicine in London 55 years ago, on November 29, 1949, Bertrand Russell had made an alarming pronouncement before the august audience when he said:

> .... We are in the middle of a race between human skill as to means and human folly as to ends. Given sufficient folly as to ends, every increase in the skill required to achieve them is to the bad. The human race has survived hitherto owing to ignorance and incompetence; but given knowledge and competence combined with folly, there can be no certainty of survival. Knowledge is power, but it is power for evil just as much as for good. It follows that, unless men increase in wisdom as much as in knowledge, increase of knowledge will be increase in sorrow.

In my own interpretation of this passage, there are two pairs of ideas conveying the central message. The first is the idea of "human skill as to means" paired with the idea of *human knowledge*; the second is "human folly as to ends" paired with the idea of *lack of human wisdom*. The central message is, of course, regarding means and ends that characterize human civilization. We employ our skills to devise various means for various purposes, and the purposes define the ends of our continued existence. Evidently, it is the *means-ends nexus* that determines human survival.

Russell is focusing on the human means-ends nexus in the present condition of humanity, with a prudential consideration of the prospect of human survival given the logic of that nexus. And, the logic of the means-ends nexus is quite clearly stated: "Knowledge and competence combined with folly [where folly is a reflection of lack of wisdom] would make human survival doubtful. Increase in knowledge without a corresponding increase in wisdom is said to be inversely proportional to happiness."

By 'means' Russell of course refers to the entire human instrumental set-up created by means of scientific-technological skill. Western technological culture has had a pervasive effect on the present human condition. The entire human situation is technologically structured, and that structure is designed apparently to make the condition of human survival more comfortable than earlier. This, in a very general sense, is the end. Human knowledge and competence are deployed in facilitating what is reckoned as human flourishing.

There may be the suspicion that unless human knowledge and competence are applied to the technological manipulation and control of the natural surrounding of humanity, humanity's survival strength may diminish in the face of adverse natural forces. There may be some truth in this suspicion. But a great deal depends on how judiciously we apply the technological skill at our disposal to our natural surrounding. It is obvious, from Russell's remark, that humanity is liable to folly, or lack of wisdom. For the human mind is capable of exercising knowledge or intelligence in a manner divorced from wisdom. When the natural surrounding is exploited by knowledge divorced from wisdom, the resultant condition is one of sorrow.

The display of marvellous, technological knowledge and the achievement of splendid technological means for bringing about spectacular changes in our living environment have been the hallmark of the last century's human image on this planet. Ironically, however, it is with utmost alarming consciousness that the same humanity has experienced, in the same century, the imminent dangers of an environment massively misused by the most knowledgeable race on earth. The so-called awareness and movement of *environmentalism* and *ecofriendliness* are direct reflections of our candid confession of eco-insensitivity. That the most intelligent species on the earth lives such an eco-unfriendly life on it is, surely, a glaring example of human folly.

One might say, using Russell's phraseology, that this awareness of environmentalism or eco-friendliness is a belated awareness of *environmental wisdom*. Environmental wisdom is necessary for our proper negotiation with the environment in our inevitable intercourse with it. Environmental ethics would therefore have to come to terms with environmental wisdom. The key question that concerns here is the question of what is the *ethical crux* of environmental wisdom.

Another way of putting the question would be whether an ethic of environmental wisdom requires a departure from human-centred morality, so that the moral significance of the environment is recognized in an ethical sphere that does not coalesce with the sphere of human morality. It is a question of whether the orbit of human morality is too narrow to include the possibility of ascription of moral worthiness to the non-human world.

The dominant ethical tradition in the western world has been human-centred, in which the core belief is that only human beings have independent moral status, and only they matter in themselves or intrinsically. The liberal philosophy of the western world holds that a human individual should be able to do what he/she wishes, providing (1) that he/she does not harm others human beings, and (2) that he/she is not likely to harm himself/herself irreparably. This may be dubbed "the Harm Principle" of western liberal ethics. Does the Harm Principle impose restriction on the freedom of human beings to deal with the environment as they please? Yes, it does. For example, you are not supposed to pollute a community on the ground that it would physically interfere with other human members of the community who use the stream. However, this does not mean that the stream is valued in itself; for no non-human beings have in this view, an intrinsic value. Only objects which are of use or concern to humans, or which are the products of human labour or ingenuity, are of value, and that too of *instrumental* value.

There are western critics of the liberal conception of ethics, who have denigrated this view as 'human chauvinism'. Such critics have advocated a kind of decentring of the status of the human individual from the sphere of moral space. Humans are not to be seen as occupying the moral centre of the world, and there is more to the moral universe than human beings. Now, the question is how far can the moral horizon extend beyond the human species. What is the range of beings—living and non-living, human and non-human—that can be credited with intrinsic moral value?

Considerable diversity of opinions exists with regard to the extension of ethical boundary beyond human chauvinism. Should it extend so as to include non-human animals, so that we can legitimately talk about animal rights and wrongs? This probably is the easiest extension because of the fact that animals are subject to pain and pleasure like humans. After all, non-human animals have a life of experiences as the humans do qualitative distinctions apart. But why stop at animals? Why not extend the moral net to encapsulate all life forms or living beings—the plant world? Isn't having a life itself sufficient for deserving intrinsic value and therefore ethical status?

Once we start widening the moral net beyond humanity and valuing reality beyond human being, we encounter difficulty in drawing a fine line of demarcation. That is, once humanity is decentred from the moral universe, we set a cosmo-centric train of thought in motion, such that more and more aspects of the entire cosmos begin to strike us as deserving intrinsic value. We leave humanity a long way behind, incorporate ecosystems, and the biosphere as a whole. The moral horizon seems to have no visible boundary. Thus, from a *shallow* human-centred morality, we move towards a morally *deep* world that hardly leaves anything out of our legitimate moral concern.

Is it that in so moving beyond human-centred morality we move beyond morality at together into some kind of cosmo-centric mysticism? It sounds as though the *deep* sense of morality with a cosmic valuedimension is more a mysticism than morality—more about unqualified, unconditioned goodness than anything falling under the ethical categories of good and bad, right and wrong. Is an environmental ethic understood in the deep sense indicated above a form of environmental mysticism?

Well, it depends on what we mean by mysticism. If by mysticism we mean that nothing of theoretical articulation is possible, then environmental ethic is not mysticism. I presume that an ethical theory with a cosmocentric stance has, just as any, cosmo-centric view has, a mystical ring to it. But it after all is a theoretical stance and hence admits of theoretical articulation. And, this articulation has to be predicated upon the justification or plausibility of the adoption of the cosmo-centric stance. In what sense are we reasonably led away from the anthropocentric to the cosmocentric value stance?

My immediate explanation required by an answer to this question rests on what I consider to be the key concept in this regard, namely, *interdependence*. The survival of anything in the universe is a matter of its vital dependence on other things, including the whole system of things, biosphere and ecosphere included. To start with, humanity's dependence on one another and on the non-human surrounding world is a clear case in point. Even if we might think that rational consciousness is a privilege that humanity enjoys in this world, that too is not our independent possession. We are as much supported by the forces of nature for its existence and continuity as our other vital functions and the same forces support their continuity. We are fully enmeshed in nature, in the cosmic order. And I suppose what goes for us also goes for other creatures available in the universe. Perhaps, this is cosmic interdependence, which I consider to be the foundation of the distribution of values over the cosmic order as a whole, without any hierarchical discrimination.

I therefore suppose that an environmental ethic draws its legitimacy from the principle of interdependence alluded to above. Since the principle itself is cosmo-centric, environmental ethics stare on a cosmo-centric axis. Quite interestingly, it is in the West that, in the later part of the last century Arne Naess, a thinker of Norway, has proposed what is well known as 'deep ecology'—a theory pretty much cosmocentric in spirit and seemingly wedded to the principle of interdependence. Naess says:

> Modern ecology has emphasized a high degree of symbiosis as a common feature in mature ecosystems, an *interdependence* for the benefit of all. It has thereby provided a cognitive basis for a sense of belonging, which was not possible earlier. Family belonging, the tie of kinship, has a material basis in perceived togetherness and cooperation. Through the extension of our understanding of the ecological context, it will ultimately be possible to develop a sense of belonging with a more expansive perspectives *ecospheric belonging* (emphasis mine).

The idea of 'ecospheric belonging' is that of interdependence of all being in the ecosphere. For Naess, this togetherness or unity with nature is a matter of becoming part of what he calls "the great self". This concept of the great self is highly reminiscent of classical Indian philosophical visions of the ultimate unity of everything. Whether in *Samkhya* or *Advaita Vedanta*, the cosmocentric vision is intended to shape our place in reality, and to regulate our conduct nof with an exclusivist principle of human autonomy, but in tune with the cosmic order.

More than in *Samkhya* or *Advaita Vedanta*, the principle of interdependence that provides the foundation of an environmental ethic is manifestly available in the philosophy of Buddhism, especially in the doctrine of 'dependent origination' or *pratityasamutpada*. That doctrine is deeply revealing of our true place in the universe, and of the value system we are to expound and espouse in our conduct in this universe.

Deep ecology would be a particular expression of the world-view implicit in this Buddhist vision of human self-understanding.

If deep ecology is a particular form in which environmental wisdom has dawned on humanity the general basis of that wisdom, an ancient wisdom indeed, is the Buddhist doctrine of '*pratityasamudpada*'.

I would like to end this note with an observation on what we humans are, endowed as we are with a distinctive feature, namely our consciousness and our rationality. Morality, as a theory, or ethics, as we know it, is a product of human consciousness and human rationality. The realization that morality must not be human-chauvinistic, or for that matter species-chauvinistic as such, whatever may be the species, is perhaps a uniquely human realization. A lion, for instance, is incapable of realizing that it is egocentric in its relation to the rest of nature. You cannot therefore condemn the lion for not being able to see beyond its lion-chauvinistic purview. It cannot decentre itself from its existential sphere. By contrast, humans would earn self-condemnation for their failure to decentre themselves and to locate themselves in the moral universe from the cosmocentric perspective. Thus, the human privilege to be the subject of rational consciousness, and to be ethical theorists, is really the privilege of being a decentred subject. The human moral subject or self is a decentred self. And it takes a decentred human self to have a proper environmental ethic. What I have called environmental wisdom is ultimately an expression of the unique way a decentred self relates itself to the universe.

4

Eco-centrism Revisited

Santosh Kumar Pal

The concern for environment is as old as our life on the earth. But, the contemporary environmental ethics has taken its formal course in the second half of twentieth century. Before that, thinkers with environmental concern placed man at absolute, objective centre of the earth. Genuine environmental ethics begins with the philosopher's approval of moral extentionism: some thinkers strongly put forward the point that our moral concern should cover non-human animals too. Darwin's theory of evolution and investigations of ecology, again, paved the way for recognition of the interdependence of living species and ecosystems, and with the side-effects of human interaction with them. Nowadays, we talk of the effects of the building up of greenhouse project, depletion of the ozone layer, deforestation, extinction of species, pollution of atmosphere and of rivers and oceans that threaten the well-being of both presently existing and future humans. Anyhow, a survey of literature on the subject reveals that although earlier proposals (at least of the West) for doing with the environment are, more or less, based on human interests, contemporary ethicists decline to confine moral standing to human interests only. They argue that some non-human animals are also sentient, and their interest in not being made to suffer has to be regarded as morally relevant. This has given rise to the theory of bio-centrism that looks beyond traditional anthropocentrism. Some contemporary thinkers, again, appeal beyond sentience to the capacity of all living organisms to develop and flourish in the manner of their own kind, and again, claim that rights belong to species or even to ecosystems. This direction of thought has culminated in the theory of eco-centrism.

The present paper is concerned with eco-centrism. In Section I of the paper I shall attempt at an exposition of the theory, and in Section II, I shall make a critical evaluation of the same. I shall show that total egalitarianism of eco-centrism is not possible, not even desirable, and that the dichotomy of *fact* and *value* is to be overcome so that a meaningful account of it could come out. In Section III, I wish to make some concluding observations. I shall argue that dualistic thinking, which has long legacy from Descartes, is to be overcome if we want to develop an adequate ecological ethic. To me, it appears that some sort of spiritualistic metaphysics, like that of Indian Vedanta, is to be acknowledge, if we want an answer to the question, why should we care for the earth.

I

Eco-centrism is that holistic theory according to which the whole ecosystem, comprising both the biotic and abiotic parts of nature, deserves moral worth. This viewpoint is based on, and inspired by, the findings of the science of ecology. The term 'ecology' derives from the Greek word '*oikos*' that literally means house. The term was coined by Ernst Haeckel, a German biologist, to denote the investigations into the interrelationships between animals, plants, and their inorganic environment. Accordingly, ecology is the systematic of all those complex interrelations as the conditions for the struggle for existence. The insight of ecology is at the root of the development of eco-centric thought.

The American philosopher, Aldo Leopold, first felt, more than half a century ago, the need for eco-centric environmental ethics, an "ethic dealing with the man's relation to the land and to the animals and plants".[1] He however, christened it as 'land ethic' that aims at enlarging "the boundaries of the community to include soils, waters, plants and animals, or collectively, the land".[2] Leopold's land ethic does not, however, prevent the alteration, management and use of these resources, but it does affirm their right to continued existence in a natural state. It virtually changes the role of *Homo sapiens* from conqueror of the land community as such. According to Leopold's prescription, "a thing is right when it tends to preserve the integrity, stability and beauty of the biotic community. It is wrong when it tends to otherwise".[3]

This holistic land ethic is the first paradigm of eco-centrism in environmental ethics. In contemporary environmental philosophy it is known as 'deep ecology' ethics. Whatever may be the title, the eco-centric ethics start from a rejection of the 'man-in-environment' image in favour of the relational, total-field image.

If we take a cursory look at the development eco-centric thought in the twentieth century, we shall find that the challenge of the traditional 'man-in-environment' paradigm has come from five main sources. *First*, the quantum theory has demonstrated that the natural world is not merely inert matter; it is alive with energy and vitality. In addition to that, this quantum theory has demonstrated that there is deep interrelation between the observer and the observed. And, as such, it is not possible for an observer to remain isolated from what is being observed. *Secondly*, the chaos theory has also denied the thesis of static equilibrium. It has drawn our attention to the fact that a minute change in the system might lead to larger repercussion in the functioning of the system, and that there is more than one direction that the system can take depending on its contextual and contingent variables. Both these two theories dislodge the traditional concept of static, mechanical nature. *Thirdly*, the theory of evolution has given a blow to the historical belief in the uniqueness of humanity. Darwin's theory not only presented an interdependent account of nature but also hinted at the strong evolutionary links between humanity and all other natural entities. *Fourthly*, the science of ecology has shown, as we already noted, that the idea of an insular, independent individual is a myth. It has demonstrated that no thing and no activity should be considered in isolation because of the circularity and holistic base of ecosystems. *Fifthly*, another source of ecological thinking may be traced back to the philosophy of Spinoza. Spinoza envisaged of a unique fusion of an integrated man/nature metaphysics with modern European science. His ethic demonstrates biospheric egalitarianism. And, science is endorsed by him as valuable primarily for contemplation of a pantheistic, sacred universe and for spiritual discipline and development.

Anyhow, by supporting such an alternative worldview, modern and contemporary science has been a crucial non-anthropocentric intellectual force in the world. Once we accept these scientific investigations and view the world in all its aspects, cycles and interrelations, then man's position would have to be seriously re-examined, as human community fails to make a firm ontological divide in the order of existence. Consequently, we feel an urge to evolve an alternative ethic based on the insights of modern science. This eco-centric theory directs us to extend our moral concern to items that are non-human, indeed to things that are not even animals, such as plants, works of art, forests, villages, families and ecosystem.

With the revival of the ecological concern in 1970s, the Norwegian philosopher Arne Naess wrote a brief but influential article in 1973 entitled, "The Shallow and the Deep, Long Range Ecology Movement". Here, he makes an important distinction between 'shallow' and 'deep' strands in the ecology movement. The former, viz., the shallow ecological thought, is limited to the traditional moral framework. Those who take the shallow strand and anxious to avoid pollution to our water supply so that we could have safe water to drink, they sought to preserve wilderness so that people could enjoy walking through it. This anthropocentric ecology is gradually overturned, and replaced by a new deep ecological philosophy. Deep ecology directs us to preserve the integrity of biosphere for its own sake, irrespective of possible benefits to humans alone that might follow from so doing.[4]

According to Naess, ecologically responsible policies are concerned only in part with pollution and resource depletion. There are deeper concerns, which touch upon the principles of diversity, complexity, autonomy, decentralization, symbiosis, egalitarianism and classlessness. Naess speaks of, more or less, seven principles in his epoch-making article just mentioned.[5] To have an adequate understanding, it is imperative to take note of these principles. These are as follows:

(i) Rejection of Anthropocentric Individualism

Deep ecology categorically rejects the 'man-in-environment' image in favour of the relational, total-field image. Naess sees organisms as knots in the biospherical net or field of intrinsic relations. An intrinsic relation between two things A and B is such that the relation belongs to definitions or basic constitutions of A and B, so that without that relation, A and B are not truly cognizable. Such a total-field model dissolves not only the 'man-in-environment' image, but every compact thing-in-milieu concept—except when talking at a superficial level of communication.[6]

(ii) Biospheric Egalitarianism

Deep ecology ethic believes in egalitarianism 'in principle'. All biotic communities, including the abiotic nature, have right to live and blossom. Naess inserts 'in principle' clause to accommodate the pragmatic aspects that "any realistic praxis necessitates some killing, exploitation and suppression".[7] The equal right to live and blossom is here seen to be an intuitively clear value axiom. Its restriction to humans is an anthropocentrism with detrimental effects even upon the life quality of humans themselves. The quality of our life partly depends upon the deep delight and contentment we obtain from the close partnership with other forms of life. The attempt to deny this fact of interdependence and the subsequent effort to dominate nature resulted into our self-estrangement.

(iii) Diversity and Symbiosis

Eco-centrism upholds the principles of diversity and symbiosis. Diversity increases the level of survival, thereby enhances novelty and richness of life forms. The proponents of deep ecology understand the Darwinian theory of evolution in positive terms. The so-called struggle for existence is understood in the sense of ability to coexist and co-operate in the complexity of relationships. "Live and Let Live" is the contention of ecology, rather than "Either You or Me". The latter principle helps to reduce the multiplicity of kinds of forms of life, supports the practice to kill, exploit and suppress. Ecologically inspired attitudes favour diversity and symbiosis of lives, of cultures, of occupations, and so on.

(iv) Anti-Class Posture

It is sometimes thought that the enhancement of quality of human life form depends on suppression and exploitation of some life forms of other groups. But ecology has taught us that is wrong: it rather leads to alienation, and in this posture both the exploiter and the exploiter are "adversely affected in their potentialities of self-realization".[8] Such an class attitude is to be altogether left out, and we have to take instead the egalitarian attitude. The anti-class egalitarian attitude inspires us to extend all the above-mentioned principles to any group including those of today between developing and developed nation.

(v) Fight against Pollution and Recourse Depletion

In the fight against pollution and recourse depletion, the ecologist finds a lot of supports. But sometimes it goes against the deep, total stand, when they focus only the pollution and depletion, without talking other aspects seriously. This takes place when, for example, projects of pollution control are implemented, but with the installation of anti-pollution devices the evils of other kind, e.g., group difference, increase. Naess says that this attitude serves only the shallow ecological stand, whereas ethic of responsibility should take into account all the principles and policies of deep ecology.[9]

(vi) Complexity, but not Complication

The theory of ecology contains an important distinction between what is complicated without any unifying principle, and what is merely complex. A multiplicity of interacting factors may operate together to form a unity or a system. Now, if we fail to take note of this aspect, we will find mere complication. Our ignorance of the biospheric interrelationship is a cause of such misunderstanding.

(vii) Local Autonomy and Decentralization

The vulnerability of a form of life is, more or less, equal to the weight of influences from outside the local region in which that form has attained an ecological equilibrium. This equilibrium lends support to our efforts to strengthen local self-government. But these efforts presuppose an impetus towards decentralization. The problem of pollution, e.g., including those of thermal pollution and re-circulation of materials leads us in this direction, as enhanced local autonomy (if we manage to keep other factors constant) might reduce energy consumption.

It should, first of all, be noted that formulation of these principles are, more or less, vague generalizations. Secondly, we have to take note that the significant tenets of deep ecology are normative and involve value priority system. Thirdly, it is also important to note that the ecological policies and the principles, as applied to our case, are "eco-philosophical rather than ecological". As Naess puts it: "Ecology is limited science which makes use of scientific methods. Philosophy is the most general forum of debate on fundamentals—descriptive as well as prescriptive—and political philosophy is one of its sub-sections. By *ecosophy* I mean a philosophy of ecological harmony or equilibrium. A philosophy is kind of *Sofia* wisdom, is openly normative."[10]

Anyhow, in a paper published in 1984, Arne Naess and George Sessions set out several principles for a deep ecology ethic, beginning with the following:

- The well-being and flourishing of human and non-human lives on earth have value in themselves. These values are independent of usefulness of non-human world for human purposes.
- Richness and diversity of life forms contribute to the realization of these values and are also values in themselves.
- Humans have no right to reduce this richness and diversity except to satisfy vital needs.[11]

Two Australians working at the deep end of environmental ethics, Richard Sylvan and Val Plum Wood, also extend their concern beyond living things, including in it an obligation not to jeopardize the well-being of natural objects or systems without good reason.[12] In 1990, Warwick Fox linked deep ecology with transpersonal psychology through the ideas of identification with the non-human world and the notion of expanded self that is capable of moving beyond.[13]

Although the contemporary ethic of eco-centrism owes much to Arne Naess, there are theorists, among others, Kenneth Goodpaster, Lawrence Johnson and Holmes Rolston III, who have contributed a lot to the propagation of eco-centric viewpoint.

Goodpaster confirms the view that being alive is the only plausible and non-arbitrary criterion of moral consideration. He hints in "On Being Morally Considerable" that this might include entities such as the bio-systems, species, etc. He maintains that a genuine environmental ethic should acknowledge Leopold's concern for the integrity, stability and beauty of the land that includes both the biotic and abiotic communities. Goodpaster refuses to accept the individualistic model of ethics that is intrinsically hostile to genuinely environmental ethic. According to him, we need to take seriously the possibility that to worthy of moral respect, a unified system need not be composed of cells and body tissue: it might be composed of humans and non-human animals, plants and bacteria.[14]

But Johnson contends that Goodpaster's position still fails to discard the individualistic flavour. The bio-system is held worthy of moral respect in virtue of characteristic that render it similar to those individuals whose moral standing we have already recognized. This contention is substantiated in his "A Morally Deep World" where he argues that various beings other than individual organism can meaningfully be said to have interests and that these interests are morally significant. These beings obviously include species and ecosystems. He thinks that these entities that have morally significant interests are not the same as the aggregated interests of their component individual organism, and that one can assert this without introducing 'metaphysical monsters'.[15]

Rolston III thinks that although plants and other non-conscious living individuals lack a subjective life, they have an objective life of a kind that entitles them to the respect of moral agents. Living things, he says, have a "good-of-their kind" as follows: "An organism is a spontaneous, self-maintaining system, sustaining and reproducing itself, executing its program, making a way through the world. [DNA-coded information] gives the organism a telos, or end, a kind of (non-felt) goal."[16] It seems that Rolston III appeals to have a form of ethical naturalism, maintaining that, as we offer appropriate biological description of species, we simultaneously uncover the correct evaluation of them. Species are dynamic life forms. The individual is taken of type; and, from the point of view of evolutionary ecology, the species is more important. On this view, extinction is a kind of super killing, in which not just individuals but the form of individuals is destroyed.

From the above exposition we understand that eco-centrism is based on the insights of western ecology, that it advocates for egalitarianism and that it rejects anthropocentric viewpoint as thoroughly wrong.

II

Eco-centrism has, no doubt, *prima facie* attractiveness, but a critical reflection will show that there are some unavoidable difficulties inherent in this theory. Here I shall be concerned only with two philosophical issues. First, I shall reconsider the so-called egalitarian anti-anthropocentric tendency of this ecological holism. Secondly, I shall throw light on the dichotomy between *fact* and *value*, between *is* and *ought* as related to our case in the point.

According to eco-centrism, all ecosystems and species have equal right to live and blossom, and we have to acknowledge it. If this is admitted, then bacteria and humans have to be considered as equal. But, such a total egalitarianism is wrong, and not only that it goes against the *modus operandi* of nature. Total egalitarianism, according to which every form of being has an absolutely equal right, is a pure nonsense. Human should be given some sort of priority in value distribution, as morality is a matter of choice and value as humans are the evaluator and creator of value. To say the truth, some form of anthropocentric is inalienable in moral discourse. If we disregard this and are guided by the so-called egalitarianism, then there might come up a demand of returning back to the original nature condition of hunter-gatherer societies. We shall have, then, to eliminate 80 per cent of the world population. Will it not be a systematic genocide? As a matter of fact, this would amount to eco-fascism[17] that can never be admitted. K.A. Parker has rightly remarked: "Blind anthropocentrism has deplorable consequences for non-human world, but a blindly misanthropic eco-centrism is no less deplorable."[18]

Undoubtedly, human chauvinism is condemnable, but we should at least acknowledge human experience as the measure of all things. It is logically possible for us to decide that a world without humans has a value, yet it is still we who are making this decision! Our point is that in order to determine our goal, and to go forward with environmentalism, this inalienable element of anthropocentricity is to be acknowledged. As long as the valuer is a human being, it is impossible for being radically non-anthropocentric in environmental matters.

But, the admission of this element of anthropocentricity does not mean the unavoidability of human chauvinism or speciesism. What is unavoidable is that we have no alternative but to make use of human yardstick in making any type of value judgement. Human valuers may find that in all consistency they must, for example, give priority to the basic needs of non-humans over the non-basic, trivial needs humans. But, the chauvinist, on the other hand, holds that interests of humans must always be accorded priority. What is notable is that the chauvinist does not take human values as a reference point, as he admits no comparison between humans and non-humans. He values human interests only because they are humans. Such chauvinism is a seriously objectionable. But, what I am pointing to is that there is an asymmetry between humans and non-humans, which is not a product of chauvinist prejudice. And, this element of humanness is precisely that which makes any ethics possible. It is human being who could develop an ethic to include other's interests into his concern. This, indeed, is the non-contingent but substantive limitation on any attempt to develop a totally eco-centric ethic.[19]

The second point I wish to examine is related to the naturalistic grounding of eco-centrism. As is often held, naturalism is antithetical to ethics. As Moore points out, it destroys the very possibility of any ethics. Anyhow, eccentric environmental ethic is, more or less, based on the natural science of ecology that studies more facts. Hence, a problem of *fact* and *value*, or *is* and *ought*, comes up. Moore has amply demonstrated that the problem of *is* and *ought* is a central problem of moral philosophy. In our case, ecology and ethics are two disciplines. Ecology is the study of facts (of nature), whereas ethics is the study of values. As an *ought* is not simply derivable from an *is*, how can the eccentric values be derived from the ecological facts? To step from fact to value is to fall in the trap of naturalistic fallacy.

The fallacy consists in the attempt of drawing a value conclusion (an *ought*) from some premises of fact (as *is*). David Hume is the first philosopher to highlight this problematic; later Moore has formulated it elaborately in his *Principia Ethica*. Anyhow, in environmental context, the question, given the facts of nature and humanity, is how is the value or respect for the environment to be derived? The problem arises because science describes natural facts, laws and their life histories, whereas ethics prescribes human conduct and directs to adopt moral viewpoint. To join the together involves a mistake, which is technically called 'naturalistic fallacy'. Eco-centric ethic, which contends that it is morally wrong to damage the earth (an *ought*), is derived from the facts of ecology, i.e., that humans exist within an interdependent biotic community (an *is*). The crucial question to explore here is: why ought the earth be cared for?

One answer that is given from some quarter is that human beings have a positive attitude to the society and the ecosystems within which they live and blossom. Contemporary investigations of science have shown us that the natural environment is also a community to which we belong. We human beings want to survive meaningfully, and so feel obliged to preserve the ecosphere.[20] Hence, the human response (the *ought*) is founded in other biophysical realization (an *is*) and moral view.

In a similar way, Edward Goldsmith has objected to the dichotomy of *is* and *ought* by arguing that the *is* is drawn substantially from the positivistic paradigm with the insistence upon objective, non-intrinsic facts. The problem appears less significant when the valuer moves outside the paradigm within which the *is* is located, e.g., when the self is located in the larger biosphere—as to project the biosphere is to protect the self.[21] For just as the new science directs valuers to abandon the sharp dichotomy between the singular individual and the surrounding world, so too the so-called distinction between valuing subjects and value—free objects be abandoned.

A third way to work around the naturalistic fallacy is to recognize that nature has intrinsic value independent of human concern. This independent value (*ought*) is found in the physical existence (*is*) and is of the natural object which is trying to seek a good of its own. Accordingly, the percept that is right to protect the environment because of its independent intrinsic value commits no fallacy as no value (intrinsic value is not value *per se*, as much as it is axiomatic) is derived from ecological facts.[22]

Some deep ecology thinkers have taken refuge in a non-naturalistic theory of intrinsic value. They suggest that the nature has intrinsic value, independently of the subject. But, such a theory of value contradicts deep ecology's own conception of them as parts of a single whole. If it is said that non-human nature has objective value and worth, it will imply, again, that human and nature are separate.[23]

III

It seems to me, if we want to overcome such a dualism and defend eco-centrism, we have to embrace the Vedanta standpoint, where all duality is seen to be dissolved with cosmic self-realization. For an acceptable version of eco-centrism, we have to put emphasis not only on the thesis of intrinsic value, but also on a form of human self-realization, which springs out of ecological consciousness. And, it is interesting to note that Naess has ultimately invoked[24] the *Bhagvad Gita* to emphasize on this self-realization: "He whose self is harmonized by yoga sees the self abiding in all beings and all beings in self; everywhere he sees the same" (*Sarvabhutasthamatmanm sarvabhutani catmani/Iksate Yogayuktatma sarvatra samadarsana*[25]).

Through such an identification, a higher level of unity is experienced from identifying with one's nearest relatives, higher unities are created through circles of friends, local communities, tribes, compatriots, races, humanity, life and ultimately with the Supreme Whole, the World in a broader and deeper sense than usual. This identification is not merely psychological affairs, but is grounded in recognition of the metaphysical reality of interconnectedness. As Naess reiterates, this way of thinking at its maximum corresponds to that of the enlightened Yogi, who sees "the same", the *Atman*, who is not alienated from anything.[26]

When we recognize the involvement of wider wholes in our identity, an expression in the scope of our identity and in the scope of our self-love occurs. As Freya Mathews explains, this appears to be the meaning of Naess theory of self-realization, and leads to a loving and protective attitude towards the world—an extension of our loving and protective attitude to our own body, so to say.[27]

This attitude of protectiveness, based on the identification with the nature, marks the shift from an ethic of duty, grounded in the recognition of the intrinsic value of selves, to an ethic of care.

Notes

1. Aldo Leopold, 'The Land Ethic' in Hugh La Follette (ed.), *Ethics in Practice: An Anthology* (Blackwell Publisher, Oxford, 1997), p. 635.

2. Ibid.

3. Aldo Leopold, *A Sand County Almanac* (Oxford University Press, Oxford, 1949) pp. 224-25 [as quoted by Alexander Gillespie in his *International Environmental Law: Policy and Ethics* (Oxford University Press, Oxford, 1997) p. 161].

4. Cf. Peter Singer, *Practical Ethics* (Cambridge University Press, First Indian Edition, 2000), p. 280.

5. Arne Naess, 'Deep Ecology' in Carolyn Merchant (ed.), *Ecology* (Rawat Publications, Jaipur and New Delhi, 1996), p. 120.

6. Ibid., p. 121.

7. Ibid.

8. Ibid.

9. Ibid., p. 122

10. Ibid., pp. 123-24

11. Cf. Peter Singer, *Practical Ethics, op. cit.*, pp. 280-81.

12. Ibid., p. 281

13. Cf. Warwick Fox, *Toward a Transpersonal Ecology: Developing New Foundations for Environmentalism* (Sambala, Boston & London, 1990).

14. See Kate Rawles, 'Biocentrism' in *Encyclopedia of Applied Ethics,* Vol.I, (Academic Press, 1998), p. 279.

15. Ibid.

16. Holmes Rolston III, 'Values in and Duties to the Natural World' in E.R. Winkler and Coombs (eds.), *Applied Ethics* (Blackwell, 1993), p. 227.

17. Cf. S. Subbarao, *Ethics of Ecology and Environment* (Rajat Publications, New Delhi, 2001), p. 146.

18. Kelley A. Parker, 'Pragmatism and Environmental Thought' in Andrew Light and Eric Katz (eds.), *Environmental Pragmatism* (Routledge, London, 1996), p. 33.

19. Tim Hayward, *Political Theory and Ecological Value* (Polity Press, Cambridge, 1998) p. 52.

20. J.B. Callicott, 'Hume's Is/Ought Dichotomy and its Relation to Land Ethic', *Environmental Ethics*, Vol.4, 1982, p. 163.

21. Edward Goldsmith, *The Way Towards an Ecological World View* (Rider, London, 1992), p. 403 [as stated by Alexander Gillespie, *op. cit.,* p. 163].

22. Holmes Rolston III, 'Is There an Ecological Ethic', *Ethics*, Vol.85, (1974), pp. 93-103 [as stated by Alexander Gillespie, *op. cit.*, p. 164].

23. David Pepper, *Modern Environmentalism* (Routledge, London, 1996), p.50.

24. Arne Naess, 'Identification as a Source of Deep Ecological Attitudes' in Tobias (ed.), *Deep Ecology* (San Diego, Avant Books, 1985), p. 260 [as stated by Freya Mathews in his paper, 'Value in Nature and Meaning in Life' in Robert Elliot (ed.), *Environmental Ethics* (Oxford University Press, New York, 1998), p. 142].

25. Subodh Majumder (ed.), *Shrimadbhagwadgita* (Deb Sahitya Kutir, Kolkata, 1986), p. 138.

26. Arne Naess, 'Identification as a Source of Deep Ecological Attitude', *op. cit.*, p. 363.

27. Freya Mathews, 'Value in Nature and Meaning in Life', *op. cit.*, p. 144.

5

Peter Singer's Views on Environmental Ethics and the Expanding Moral Community

Ramdas Sirkar

Peter Singer is one of the most celebrated philosophers of our time. As an influential leader of the practical ethics movement, he developed an environmental ethic. Environmental ethics is the study of normative issues and principles concerning the man-nature interaction.

In the first section of the present paper, I shall state briefly Singer's central idea of a non-speciesist ethics. His theory of environmental ethics will be explained in the second section. In the last section, I shall discuss his critique of the ethics based on the fundamental moral attitude of respect for nature shall be discusses.

I

Singer's views on ethics recognizes an important role of reason in our ethical discussion. This role of reason is evidenced by our admission that the very idea of living according to ethical standards is closely related to our ability to justify the way in which we live. He notes that ethical justification cannot be given in terms of self-interest only. It must extend to the interests of others. This universal aspect of ethics, he suggests, provides a 'persuasive' ground for accepting a broadly 'utilitarian position'. Thus, if I am to be ethically concerned for my own interest, then I must extend this concern to the like interest of others. This leads to the maxim that we "must choose the course of action which has the best consequences, on balance, for all affected".

Some philosophers argue that in our ethical judgement we must consider interest simply as interest. Now, if it is accepted, Singer claims, we can formulate the basic principle of equality "as equal consideration of interests is that we give equal weight in our moral deliberations to the like interests of all those affected by our action". He argues that this principle extends beyond the human species. For, once this principle is recognized as a moral basis for relations among the members of our own species, we are bound to recognize the same principle as a moral basis for our relations with the members of other species, i.e., the non-human animals. He contends that if it is morally wrong to disregard interests of some members of our species on account of their race, sex on intelligence, then it is also morally wrong to disregard the like interests of the members of animal species on account of their being less rational. So, we cannot exploit animals just because they do not belong to our own species.

Following Jeremy Bentham, Singer maintains that the capacity for suffering or enjoyment is the necessary condition for having an interest at all. Thus, whatever be the nature of being, this principle of equality demands that their suffering be considered equally with the like sufferings of other beings. To quote Singer, "the limits of sentience are the only defensible boundary of concern for the interests of others". Therefore, in our ethical deliberations, we must take account of the interests of all sentient creatures, humans and non-humans, self-conscious and non-self-conscious. By granting moral consideration to non-humans, Singer provides a non-speciesist ethics.

II

In the second edition of his book *Practical Ethics*, Singer begins his enquiry into the possibility of environmental ethics by noting that the traditional western ethical thought presupposes a particular attitude to nature. This particular western attitude to nature has evolved from the biblical account of creation. In the *Genesis* we find the Hebrew view of the special status of human beings. There, human beings were granted dominion over nature. During its Roman period, Christianity absorbed ancient Greek ideas of nature and particularly that of Aristotle. Aristotle viewed nature as hierarchical. For him, human beings having less rationality exist for the sake of those having more. Thus, in the mainstream Christian thought, the possibility of sinning against non-human animals or against the natural world is ruled out. Presently, thinkers debate the interpretation of this grant of 'dominion'. John Passmore and some other philosophers try to read into this grant a directive to act as stewards, i.e., to look after nature on behalf of God. However, Singer finds no justification for such an interpretation of the text.

The dominant western tradition, influenced by the mainstream Christianity, assumes that the natural world exist for the benefit of humans. Morality begins and ends with human beings. Nature being devoid of any intrinsic value, the destruction of any part of nature is not sinful, if by such destruction we do not harm other human beings. Thus, for two thousand years, western ethics has been anthropocentric. However, Singer suggests that within the moral framework of this human-centred western tradition, it is possible to develop environmental values like the preservation of nature. This, he thinks, can be done by relating our concern for nature to human well-being, of present and future generations.

Singer is convinced that even within the anthropocentric moral framework, "the preservation of our environment is a value of the greatest possible importance". For example, he draws our attention to the fact that the greenhouse effect threatens to cause a rise in sea level that will inundate the low-lying coastal areas like the Nile delta in Egypt and the Bengal delta, affecting the house and livelihood of 46 million people. This calls for our serious concern about environmental preservation.

The argument takes a stronger form if we formulate it in terms of future generations. When a virgin forest is cut or drowned to build a dam, the link with the past and the natural lifecycles of the plants and animals are destroyed. Obviously, no short-term benefit can "buy back the link with the past represented by the forest". Thus, he recognizes the priceless and timeless value of wilderness. However, he is quick to note that his argument does not show that cutting forests cannot be justified in special cases. In such special cases, the justification should take full account of the value of the forests to the future generations. He also accepts the argument for preservation based on appreciation of the beauties of the wilderness. He points out that for many people, the wilderness is the source of the aesthetic feeling having spiritual intensity. He even urges us to encourage future generations to have a feeling for nature. Again, it is a unique experience to see a part of nature that is untouched by human being. We cannot deprive future generations in this regard. By destroying wilderness, we shall be causing irreparable losses on the generation to come.

This rarity of the wilderness provides another strong argument for the preservation of wilderness and its irreplaceable resources even within the anthropocentric moral framework. We have seen that Singer's ethical theory proposes an extension of the ethic of dominant western tradition. He also claims that a truly environmental ethic can be developed on the basis of such ethical extension. To quote him: "At its most fundamental level such an ethic fosters considerations for the interests of all sentient creatures, including subsequent generations stretching into the far future. It is accompanied by an aesthetic appreciation for wild places and unspoiled nature."

Obviously, as a practical ethicist, Singer cannot disregard the growing environmental concern. Yet, for him, a tree, a mountain or a rock does not have interests and as such, cannot be harmed by human action. Hence, they can have no place in ethical discourse. Thus, he felt the need to make room for environmental value in his moral philosophy. He found the way out in the concept of *habitat*. He noted that the stream of mountain was the home of many animals whose rights should be respected. On this ground the destruction of environment by human action could be unethical.

In any serious exploration of environmental values a central issue will be the loci of intrinsic value, i.e., valuable in itself. Opposed to this value is instrumental value, i.e., value, as a means to other ends. As a non-speciesist moral philosopher, Singer argues that if we find intrinsic value in human experiences, we cannot deny this value in at least some experiences of non-human beings. For him, intrinsic value extends up to sentient creatures. So, if a proposed dam would cause suffering or even kill thousand or more of sentient creatures, by inundating the river valley, then in our cost-benefit analysis, we must take account of this loss. Moreover, if we destroy the habitat of sentient creatures by building the dam, then the loss would be a continuing one.

It is a truism for Singer that a society's ethic must consider those conditions that are necessary for our survival and also for a stable and lasting community. Presently, the rapid increase in population and the environmental pollution caused by growth industry threaten to wipe out our society. Though the danger of our environment is not imminent and obvious, we should develop a sensible environmental ethic within a short period. This ethic, he suggests, would regard every environmentally harmful action as questionable and unnecessarily harmful actions as ethically wrong. Thus, saving and recycling would be regarded as virtue, while unnecessary consumption would be regarded as a vice. Even our preference for any particular type of recreation is not ethically neutral. For example, the additional consumption of fossil fuel and the consequent discharge of carbon dioxide makes motor car racing ethically less acceptable than cycling.

Considering the plight of the people who live in cities and towns, this ethic would encourage us to keep our families small. In the present industrialized societies accumulation of consumer goods provides the yardstick of success. The environmental ethic developed here would not approve of such materialist ideals. Rather, this ethic would measure success in terms of development of one's ability and of experiencing real fulfilment and satisfaction. This ethic also fosters frugality for mitigating environmental pollution. Peter Singer even goes to the extent of saying that wasting of materials that can be recycled is "theft of our common property in the resources of the world". By practising frugality, we shall enter into a different type of moral life where consumption of unnecessary product will be considered morally wrong.

This new ethic demands that we be more thorough in our understanding of extravagance. The world we live in is now under pressure. In this context, if we consider the long-term value of the rainforest, then the timber products made by destroying this forest are extravagant. Similarly, the paper products are also extravagant, because these are produced by way of destroying ancient hardwood forests.

Extravagance in our food habits is ultimately a matter of serious environmental concern. Singer claims, 38 per cent of grain crop produced all over the world is now used as animal foods. The number of domestic animals is three times more than the human population. World's 1.28 billion cattle outweigh the human population.

The factory farming method adopted by the industrialized countries is responsible for the huge consumption of fossil fuels. Chemical fertilizers and farm animals produce the greenhouse gas, viz., nitrous oxide. Forest dwellers, both human and non-human, are driven out of their homeland in order to clear the forest for the grazing of cattle. In course of this grazing huge quantity of carbon dioxide is released in the atmosphere. The world's cattle are also thought to produce 20 per cent of the methane released into the atmosphere. All these constitute a compelling reason for plant-based diet. Though this environmental ethic encourages simple life, it never condemns pleasures. Of course, these pleasures do not come from over-consumption. Instead, we can find real pleasure and satisfaction from warm personal relationship, from being close to our children and friends. Environment-friendly sports and recreation can also be a source of such pleasures. We can have enough pleasures from plant-based diet. Appreciation of the beauty of wilderness is another source of such pleasure.

III

The type of non-anthropocentric individualism exposed in Singer's non-speciesist ethics has, at least, two presuppositions. First, it presupposes that only an individual can be the bearer of interests. This is so because only an individual organism can possess sentience. Secondly, this ethical view also assumes that an entity is real if an only if it is observable or if it can be confronted. Acceptance of these two assumptions leads Singer to exclude all types of holistic entities like species, ecosystem, etc., from the moral community.

Singer examines two such attempts to extend the boundary of moral community. The first attempt he considers is an ethic based on the attitude of reverence for life as advocated by A. Schweitzer and Paul Taylor. Though sharing both the assumptions they contend that moral consideration does not end with sentience. Schweitzer considered every organism as individuals whose suffering or death must be avoided if possible. According to Schweitzer, "just as in my will-to-live there is a yearning for more life" and an exaltation called pleasure and also fear annihilation, so is the case with "the will-to-live-around me", "a man is really ethical only when he obeys the constraint laid on him to help all life" when possible. In his opinion, such an ethical person does not shatter even an "ice crystal that sparkles in the sun". Singer has rightly pointed out that Schweitzer's position is not acceptable because an ice crystal is not alive at all.

Taylor's more refined view is that every living thing is "pursuing its own good in its own unique way". This helps us to see all living things as we see ourselves. Therefore, the existence of these has the same value as our own existence. To say that all living things have a good of their own is simply to say that it can be benefited or harmed. This good is objective in the sense that it is independent of what any conscious being happens to think about it. For example, a certain quantity of water is good for a particular plant whether I acknowledge it or not. Again, I can claim that this water is good for that plant without supposing that the plant itself knows this. Possessing such 'good-of-its-own' is necessary for deserving moral respect. Now, Taylor believes that every living being that has a good-of-its-own merits moral consideration. He also upholds the view that the realization of the good of an individual is intrinsically valuable. These basic ideas constitute the fundamental moral attitude that Taylor calls respect for nature. Singer finds difficulty in the defenses provided by Schweitzer and Taylor. The difficulty is due to the metaphorical use of language. We often talk about plants 'seeking' water so that they can survive. This often helps us to talk about their 'will-to-live' or of their 'pursuing their own good'. Singer points out that as plants are not conscious and cannot engage in intentional behaviour, this use of language must be metaphorical. Singer's objection runs like this: "One might just as well say that a river is pursuing its own good and striving to reach the sea, or that the 'good' of a guided missile is to blow itself up along with its target." He points out that plants never experience 'yearning', 'exaltation', 'pleasure' and 'terror'.

He again claims that it is possible for us to offer a purely physical explanation of the behaviour of plants, rivers and guided missiles. So, in the absence of consciousness, there is no good reason for having greater respect for the physical processes that take place in the growth and decay of living things than for these in the non-living things. Though there may be some difficulties in Taylor's ethical theory, Singer's critique of this theory does not appear to be sound enough. His grouping plants, rivers and guided missiles in the same class, evidence this. In spite of the fact that plants lack conscious will and intentional pursuit, there are important differences between plants, rivers and guided missiles which Singer fails to see. He considers all these as purely physical processes.

In this context, it is important how Holmes Rolston III distinguishes a plant from other inanimate objects. First, though not an experiencing subject, a plant is not an inanimate object either. It is not a geomorphological process like a river. Plants are alive and self-actualizing—it can produce vegetative modules as well as reproductive modules. Secondly, like any other organism, sentient or not, a plant, Rolston says, "is a spontaneous, self-maintaining system, sustaining and reproducing itself, executing its programme, making a way through the world ..."

Something more than physical causes, even when less that sentience, is operating within every organism. There is *information* superintending the causes: without it the organism would collapse into a sand heap... it gives the organism a *telos*, 'end', a kind of (non-felt) goal, this information is coded in the D.N.A. Thus, for him, the genetic set is a normative set. So, an organism is an evaluative system. We "pass to value when we recognize that the genetic set is a normative set". Though the organisms have no will or desires, they have their own standards. Any organisms have a good-of-its-kind and it defends this. Hence, when Taylor claims that a plant has a good-of-its-own, this claim cannot be dismissed as metaphor.

Finally, we must admit that Peter Singer has enlarged our vision of ethics, which was too humanist in the past. He is very much responsible for the change of our attitude towards animals. In spite of this advance, we must take note of the important insight provided by the ultimate moral attitude that Taylor calls respects for nature.

6

Does Anthropocentric View Rest on a Mistake?

Bibhu Prasan Patra

In this paper, an attempt is made to critically examine the anthropocentric view of the relationship between nature and human beings. The human-centred view evaluates nature (wild animals, trees, mountains, rivers, lakes and places of natural beauty) on the ground that the basic needs of humans like space, food, shelter, oxygen, water, energy, etc., are provided by nature for their sustenance. Objects of nature merely exist for the use of human beings and are valued because they serve as instruments for the satisfaction of human needs and desires.

The anthropocentric view presupposes man's superiority over nature, on the ground that man is the only rational being under the sun. This presupposition dominates the attitude of human beings towards nature and leads to the view that whatever is naturally given is only for the sake of rational members. The philosophical thinking behind this assumption (that humans are the only rational being) can be found in the writings not only of Augustine and Aquinas but also of Descartes and Kant. Descartes eulogized the rational capacity of man. He distinguishes between thought and extension and holds that thinking can only be ascribed to human being. He claims that other animals do not have sensation and feelings.

The hypothesis that animals do not even possess feelings and consciousness, shaped the ethical perspective of humans in such a way that non-humans are kept apart from our moral considerations. But, can we really deny feelings and consciousness to animals? It will be very difficult to do so, because animals exhibit their consciousness while courting, mating and caring for their young ones. They flock together to face danger. It is argued that animals do these only instinctively. But it is also true that they are conscious about these instinctive feelings. Kant very emphatically points out that humans are the only being on earth that possess understanding. He writes:

> He (the human being) is certainly the titular lord of nature and, supposing we regard nature as a teleological system he is born to be its ultimate end.

Of course, the defenders of the rationality thesis will say that lower animals may be conscious but they are not self-conscious. Self-consciousness implies the ability to distinguish experiences that one regards as aspects of one's being from those that are not.

Sprigge (1984) argues that in some form or other animals also experience aspects of one's own being, which is not one. According to him,

> ...the consciousness of a cat includes elements which it experiences as self which stand in contrast to there elements which it as not-self that there is a way in which the cat experiences its own body, its internal pleasure and pains, the warmth on its and the other cat it attacks. It may be that there are forms of animal life, at a rather low level, where there is just consciousness of a world which is not divided into self and non-self, but any animal which is related to the environment in quite a complex perceptual way must surely experience a contrast between what is itself and what is not, though obviously it does not have verbal expressions for this contrast.[1]

The above discussion presented by Sprigge shows that, in an extended sense, other beings also have the capacity to distinguish between the self and others. Human beings, only because of their superiority complex, treat others as means to their own welfare. The purpose of everything is to serve themselves. But other living beings have equal right to enjoy the facilities available in the natural ecosystem. They do not exist merely to be used by human beings; rather, they have values of their own. The instrumental value we ascribe to other things is indicative of the selfishness of humans. For the betterment of their own species, humans consider other things and beings to be valuable. Thus, the anthropocentric view provides utilitarian approach to nature, which lacks the virtue of respecting nature for its own sake. The utilitarian ethical consideration that prevails in human interpersonal transaction is reflected in a crude form in this version of the utilitarian consideration of nature.

The purely egoistic hedonism of Bentham holds that "pleasure is the only good". Humans should do that which gives pleasure to them. This egoism is extended further while considering nature. That means the pleasure of human beings only should be given utmost importance. Other living and non-living things are there only for human pleasure. The cost and benefit of everything else is measured by human welfare. The cost here is pain and benefit is pleasure. If the natural ecosystem is used for human welfare and produces pleasure in man, then there is a need to protect the environment. The refined version of utilitarianism presented by J.S. Mill emphasizes "maximum pleasure for the maximum number". But, here again, the ethical considerations are limited to maximization of pleasure or happiness for the maximum number of human beings. We are not only concerned about the present generation of human beings, but also for our future generations.

Why is it that our ethical concern is so narrow? Why is it difficult to bring other living and non-living things into our moral framework? Perhaps, our conception of feeling pleasure/happiness is restricted only to human animals.

Other lower animals are not capable of consciously measuring their pleasure and pain. As we have already discussed, in an extended sense, other creatures have these feelings but they are unable to express it. We humans are so prejudiced about our own species that we do not care for the needs and interest of other species.

Peter Singer explains this idea as follows:

...that it is anthropocentric, even species, to order the value of different lives in a hierarchical manner. If we do so we shall, inevitably, be placing ourselves at a top and other beings close to us in proportion to the resemblance between them and ourselves. Instead we should recognize that from the point of view of different beings themselves, each life is of equal value. It may be that a person's life may include the study of philosophy while a mouse's life cannot; but the pleasure of a mouse's life is all that the mouse has, and can be presumed to mean as much to the mouse as the pleasures of a person's life mean to the person. We cannot say that the one is more or less valuable than the other.[2]

But Singer is a bit doubtful about how one would really compare the value of different lives. There is a practical difficulty in finding a neutral ground. Thus, he insists that in practice one has to accept a hierarchy of values. He argues:

> The idea of a hierarchy of value leaves open the possibility that when I weed my vegetable garden, the life of each weed I destroy has some value, though a value overridden by my own needs; but is there really any intrinsic value at all in the life of a weed? Suppose that we apply the test of imagining living the life of the weed I am about to pull out of my garden. I then have to imagine living a life with no conscious experiences at all. Such a life is a complete blank; I would not in the least regret the shortcoming of this subjectivity barren from existence. This test suggests therefore that the life of a being that has no conscious experiences is of no intrinsic value.[3]

This argument shows that maintaining a hierarchy of values is essential for human existence. Human beings draw their sustenance from nature and value the natural ecosystem in accordance with their needs and desires. If at all we value other beings and the natural surroundings, we do so only in the sense of an instrumental value. Other things and beings are valued only in terms of their usefulness to human individuals.

But I want to point out here that if the utilitarian consideration dominates human thinking, then proper respect for other creature will not arise. And, so long as man's relationship to nature is valued by considering nature to be our utilitarian source, there will be no place for any ethical consideration of the natural ecosystem. It can only be used and transformed by human wish and will. The Baconian, Cartesian doctrine glorifies humans as conqueror of nature. They can manipulate and improve upon these natural surroundings, demonstrating that they are superior to non-humans.

It is again the Darwinian theory of evolution, which claims humans to be superior to non-humans. Darwin's theory emphatically states that the human species have emerged as the fittest in the struggle among all the species. But, the deep ecologists refute this by saying that the Darwinian theory can also be cited as a strong ground for equal treatment of all beings in the earth's community life. The human superiority can be countered by the fact that, before the human species came into existence, other species were already there. If human beings claim that they have emerged as the stronger species, and have the right to enjoy nature, then it is a wrong view to espouse. It is wrong because it would be very difficult to prove whether the emergence of man as the strongest species was possible because of the conscious effort of the species, or it is only because of a chance factor. The Darwinian principle of "survival of the fittest" may certainly be a sound and plausible scientific theory, but by itself the theory has nothing to show the alleged superiority of the fittest, nor did Darwin himself have anything to suggest to this effect. The 'is'-'ought' distinction is only too obvious to have escaped his reasoning. Assuming that, as a matter of fact, a certain species is the fittest in Darwin's sense, it does not follow that members of this species ought to be superior to the rest of the species. It is rather common knowledge among philosophers that a scientific explanation about the nature of species cannot, by itself, justify a moral theory about how any particular species is to be treated vis-à-vis other species. The alleged superiority of the human species, despite their 'superior' fitness, therefore remains very dubious.

What I am trying to point out, on the contrary, is that everything in the universe is a part of the natural process. And, this natural process is something that is beyond the control of any species. This assumption provides a cosmic vision that promotes the thinking that humans are not the controlling authority of natural happenings and will give different picture of the nature-human relationship. Once the thinking process is elevated to this broad vision (that humans are just a part of the nature), a strong foundation for environmental ethics will be laid. Because, once you accept that you are part of an integral whole, you will also be able to develop a sense of sharing with the others. The deep ecologist in fact looks for a foundation, which is based on this type of thinking.

The deep ecologists further argue that the controversy exists not only between the human and the non-human species, but also within the human species (i.e., between the so-called developed or industrialized countries and the developing or underdeveloped countries); indeed, the former is only a symptom of the latter. The developed nations of the world, because of their material progress and well-being, treat other nations in a very different way. The way the industrially developed nations deplete natural resources and dump the toxic waste show their contemptuous attitude to human beings living in the Third World. If we take it for granted for a moment that the anthropocentric view is right because it argues for the human species and everything in favour of human needs and interest, the attitude of the so-called developed countries puts a big question mark to this idea. It would not be out of place to mention here that leaving aside the distinction between humans and non-humans, there is intra-human discrimination that is equally deplorable. The root cause of all the discriminations (whether it is between humans, non-humans, or between humans and humans) lies only in the greed of humans. It is the greed of human beings, which led to all kinds of malpractice not only with nature but also with fellow human beings. However, we do not have scope here to discuss in detail how human beings of the Third World countries are treated by their fellow human being of the so-called developed countries. Mere mention of this should serve the present purpose adequately.

If we are seriously looking for a better community life that is shared by both humans and non-humans, then we have to broaden our vision and limit our greed. This, in fact, will provide a strong foundation for environmental protection. Now, the question is, how to achieve this goal? What kind of philosophical/metaphysical view would set such a foundation? The thinking process has to be changed in order to incorporate nature into our moral framework. The sense of sharing will be strengthened only if we (the humans) accept that we are a part of the integral whole.

II

The deep ecological perspective presented above certainly provides a sound justification for environment protection. This view is more fundamental and rational than the mere anthropocentric viewpoint. The argument they present is not utilitarian. They argue for the preservation of the nature not for mere conservation. Their emphasis is on the intrinsic value of nature. But, my contention is that this standpoint also lacks a strong philosophical/metaphysical backup. And, without a strong philosophical, metaphysical support, it would be difficult to develop a deed respect for nature.

In the western philosophical tradition, Spinoza's pantheism and Leibniz's monadology depict nature as manifestations of God. But, this view fails to guide human behaviour in so far as dealing with nature is concerned, because philosophical thinking of this type remains only at the level of abstract speculation. It has not influenced much the worldview of the western tradition. On the other hand, the Baconian thought has dominated the western thinking so much that they fail to look beyond the scientific temperament and always get themselves busy in unraveling the secrets of nature. Although the deep ecologists argue for "value of nature for its own sake", they do not provide a strong foundation for an environmental ethics.

The Indian philosophical tradition provides a solid foundation for adequate concern and deep respect for nature. Philosophical thinking in

India has always been associated with integral practice. The philosophical speculations that were developed in the Indian tradition were deeply concerned with life in general. Its aim was not just to understand nature and satisfy intellectual curiosity. Rather, the theoretical speculative thinking was associated with a strong insight that was guiding the ethical/theological aspect of human life. In fact, the philosophical ideals the Indian seers were practised in their daily life. The motto of their philosophizing has always been: *"mansyekam, vacasyekam, karmanyekam, mahatmanah"*.

The cosmic vision of the Indian philosophical thinkers depicts nature as fundamental to any existence. Nature is conceived as an entity from which everything has evolved. Human beings are considered only as a part of nature. Nature is not considered merely as a physical world, which is external to man or alienated from humans, nor are humans considered as essentially spiritual and alien to nature.

In the Samkhya system, *prakriti*[4] or the *mula prakriti* is the material cause of every existence. It is the vital source of every being and becoming. Human beings as conscious beings are conditioned by the natural world. The interaction between physical/natural world and humans is a logical necessity. *Prakriti* is helpful for a better manifestation of consciousness.

In the Samkhya metaphysics, *prakriti* has been depicted as the source of the entire process of evolution. The *gunas*, which constitute the *prakriti*, are *sattva* (pleasure), *rajas* (pain) and *tamas* (indifference). These three *gunas* pervade the entire being and evolve into the world of diversity because of the proximity of reflection of *purusa* (consciousness). The evolution is from unity to diversity.[5] In this framework, human beings are seen as one of the evolutes along with the rest of creation. Since the effect (*karya*) is real in the cause (*karma*), the world of multiplicity is contained in the original *prakriti*. Every related thing or being therefore must have three *gunas* as their constituent.

Therefore, the difference between one object and another is one of degree not of kind. The distinctive character of a particular animate or inanimate depends on the relative preponderance of a particular *guna* over the rest.

This framework leaves no room for discriminatory treatment and provides a secure foundation or a holistic living. The human species cannot have justification for thinking themselves to be the very end and non-humans as the means. The fact remains that in creation there is hierarchy of creatures depending on the degree to which consciousness is reflected. But that does not lend any rationale for the higher to live at the expense of the lower. If human beings are more evolved, then they obviously bear the obligation of protecting the interest of the non-humans. The flora and fauna are as much integral parts of nature as human beings. Therefore, humans must not alienate themselves from others, rather they should live in harmony with the rest of creation.

In the Upanishads the ultimate reality is conceived as Brahman. In spelling out the nature of Brahman and the world the Upanishads reiterate that Brahman expresses itself in the form of manifold particulars, which implies that everything and being should get equal respect in the universe. The one is not the creator of many but has become many. In this sense, the whole world is pervaded by Brahman.[5] Every manifestation—animate/inanimate, mobile/immobile—contains the spark of the divine. There is an essential unity of all existence. Diversity is only phenomenal, unity is real. The Upanishads proclaim that everything has come from Brahman and returns to Brahman. Everything is noble and holy. Since everything is essentially divine, the ideal ethical code espoused in the Upanishads is that true enjoyment consists in renunciation and not in acquisition.[7]

Thus, the Upanishadic philosophy provides the human race a worldview, which says everything and being should get equal respect in the universe. There is no hierarchy among the existent beings. The idea that divine is all pervasive reminds us not to have any bias or prejudice against the other species of the universe. It precludes the idea that other things and beings only have instrumental value for human beings. The cosmic vision of the Upanishads considers all objects of the universe—animate/inanimate, humans/non-humans—as part of the one impersonal reality (*sat*). Sri Aurobindo, in his philosophy of the Upanishads, points out:

> The existence of some Oneness, which gives order and stability to the multitudinous stair of visible world, the Aryan thinkers were from the first disposed to envisage and they sought painfully to arrive at the knowledge of that Oneness.[8]

The universe is one unit, there is no hierarchy and there is no supremacy of one species over the other. One reality is asserted, i.e., "All is God" (*"Sarvam Khalu Idam Brahma"*). There is no multiplicity here (*"neha na nasti kincana"*). Everything rest in this impersonal reality and returned into it at the time of dissolution.

The metaphysical outlook, which I have just presented, asserts that there is an eternal moral order in the universe. In the *Rig-Veda* this eternal moral order is described as the *Rita*. *Rita* is an unseen principle that sustains the entire universe with a moral principle. It is the basic law by which all living and non-living beings are regulated. The belief in the eternal moral nature of the universe shows that man is part of nature and governed by the basic laws of nature. Ethics is thus an integral part of the metaphysical/speculative thinking of the Indian philosophical tradition. A deep respect for nature was inherently present in its philosophical presuppositions. This mode of thinking has its inevitable bearing on human beings' relation with nature.

I want to point out here that the deep ecological perspective of the western tradition in recent years needs to be supported by a metaphysical worldview. Without such a metaphysical perspective, it will be very difficult to justify their standpoint and come out of the Cartesian egocentric predicament. The ethical theories that are developed in the West, whether it is teleological or deontological, fail to guide action and nurture a deep respect for nature. In so far as the duties, responsibilities and obligations of humans to nature are is concerned, a change in human attitude towards nature is needed. Despite its immediate utility and glamour, anthropocentric morality is fraught with serious limitations.

Notes

1. Sprigge, T.L.S., "Non-Human Rights: An Idealist Perspective", *Inquiry*, 27, 1984.
2. Singer Peter, *Practical Ethics*, Cambridge University Press, 1979, pp. 88-89.
3. Ibid, p. 92.
4. *Prakriti is* not the exact equivalent to the English word 'nature'. It is considered to be an eternal unconscious principle.
5. Just as the oil and the wick cooperate to produce light, the three *gunas* of the *prakriti* cooperate with each other for the creation and continuance of anything and being in the universe.
6. *"Ishavasyam Idam Sarvam"*, *Isha Upanishad*, verse I.
7. *"Tena Tyaktena Bhunjitha"*, Ibid.
8. Aurobindo, *Philosophy of the Upanishad*, p. 6.

References

Attfield, Robin (1983), *The Ethics of Environment Concern*, New York: Columbia University Press.

Baier, Annette (1984), "For the Sake of Future Generations", in T. Regan (ed.), *Earthbound: New Introductory Essays in Environmental Ethics*.

Blackstone, William T. (1974), *Philosophy and Environment Crisis*, Athens, Ga: University of Georgia Press.

— (1980), "The Search for an Environmental Ethic", T. Regan (ed.), *Matters of Life and Death: New Introductory Essays in Moral Philosophy*, New York: Random House.

Callicot, J. Baird (1984), "Non-Anthropocentric Value Theory and Environmental Ethics", *American Philosophical Quarterly*, 21: 299-309.

Devall, Bill, and George Session (1985), Deep Ecology: *Living As If Nature Mattered*, Layton, Utah: Gibbs M. Smith, Inc.

Enrenfeld, David W. (1972), *Conserving Life on Earth*, New York: Oxford University Press.

— (1978), *The Arrogance of Humanism*, New York: Oxford University Press.

Fankenal William K. (1977), "Ethics and the Environment", in K.E. Goodpaster and K.M. Sayre (eds.), *Ethics and Problems of the 21st Century*, Nortre Dame, Ind.: University of Notre Dame Press.

Godfrey-Smith, William (1979), "The Value of Wilderness", *Environmental Ethics*, Winter: 309-19.

— (1979), "From Egoism to Environmentalism" in K.E. Goodpaster and K.M. Sayre (eds.), op.cit.

Goodpaster, Kenneth E. and K.M. Sayre (eds.) (1979), *Ethics and Problems of 21st Century*. Notre Dame, Ind.: University of Notre Dame Press.

Hargrove, Evgene C. (1989), *Foundations of Environmental Ethics*, Englewood Cliffs, Prentice Hall.

Lawrence E. John (1991), *A Morally Deep World: An Essay on Moral Significance and Environmental Ethics*, Cambridge University Press.

Miller Peter (1989), "Descrartes" Legacy and Deep Ecology", *Dialogue*, Vol.28, November, 2:183-202.

Naess, Arne (1973), "The Shallow and the Deep, Long-Range Ecology movements: A Summary", *Inquiry*, 16: 95-100.

Norton Bryan G. and Henry Shue, (eds.) (1986), *The Preservation of Species*, Princeton, N.J.: Princeton University Press.

Patridge, Ernest, (ed.) (1981), *Responsibilities to Future Generations: Environmental Ethics*, Buffalo, N.Y.: Prometheus Books.

Passmore, John (1974), *Man's Responsibility for Nature: Ecological Problems and Western Traditions*, New York: Charles Scribner's Sons.

Regan Donald H. (1986), "Duties of Preservation" in Bryan G. Norton and Henry Shue, (eds.), *The Preservation of Species*, N.J. Princeton University Press.

Regan Tom (1982), *All That Dwell Therein: Animal Rights and Environmental Ethics*, Berkeley and Los Angeles: University of California Press.

Rodman, John (1977), "The Liberation of Nature"? *Inquiry* 20, Spring: pp. 83-145.

Rolston Holmes III (1975), "Is there an Ecological Ethic"? *Ethics*, 85: pp. 93-109.

— (1994) Conserving Natural Value, New York, Columbia University Press.

— (1988) *Environmental Ethics: Duties and Values in the Natural World*, Temple University Press.

Salthe M. Barbara and Stanley M. (1989), "Ecosystem Moral Considerability: A Reply to Cache", *Environmental Ethics*, Vol.II, No.4: 355-61.

Singer, P. (1979), *Practical Ethics*, Cambridge University Press, Cambridge.

Sprigge, T.L.S., "Non-Human Rights: An Idealist Perspective", *Inquiry*, 27, 1984: 439-61.

Taylor, Paul (1986), *Respect for Nature: A Theory of Environmental Ethics*, Princeton University Press.

7

Tracing the (Dis)continuities: Heidegger on Technology

A. Raghuramaraju

Martin Heidegger both exalts as well as criticizes the western universalism. Along with presenting a critique of technology, he also wonderfully sums up the journey of western metaphysics that is rooted in 'subjectivist-metaphysics'. His preoccupation with 'Being' explicates as well as critiques the western metaphysical thinking—the enabling condition of technological culture. In this explication of 'Being', he is an insider; this is more prevalent in his earlier writings particularly in his magnum opus *Being and Time*. However, one can decipher some shades of his later critical attitude towards western thought even in his earlier writings. This paper focuses on his critique of metaphysics, lay bare those aspects of continuity within the West.

Heidegger locates the rise of technology, which results in ecological crisis, in these continuities. In this paper, while accepting his critique, I however contest his portrayal of continuities in the West from the classical to the present. By focusing on these sharp discontinuities within the West, I maintain that he overlooks these significant discontinuities or merely subsumes them under continuities. In what follows, I shall present one major discontinuity, namely, modern technology that eludes Heidegger. I shall locate the emergence of modern technology in the dialectic of these continuities and discontinuities.

Before explicating this and by way of making a pace for it, I would in the following elucidate some broad perspectives on technology and ecology. There are those who treat technology as value-neutral thus making its use and misuse either user-specific (Duchet) or class-specific (Marx, Mounier, Marcuse in Durbin 1979). This view reduces technology to a mere instrumentality. There are those who treat technology as value-embedded, trace its rise to anthropocentrism. However, there are differences in this group regarding the origins of anthropocentrism.

(i) For Heidegger, it is Plato who by making the disclosure of Forms (Absolute Being) dependent upon how the viewer perceives those Forms made subjectivist-metaphysics possible. This privileges technological culture, causing ecological crisis. He goes to the extent of equating western philosophy (metaphysics) to subjectivist-metaphysics, which is all pervasive. This mode of thinking inaugurated by Plato culminates in Nietzsche's nihilism. Even the Cartesian subjectivism is possible within the framework initiated by Plato. Heidegger's "avoidance of traditional categories of metaphysics" deriving "from the substance-centred as well as subject-centred metaphysics of the western philosophical traditions" is very suggestive of the extent and pervasiveness of the subjectivist-metaphysics. This western metaphysics, culminating in the scientific and technological mode of thinking dominating the man of today, has assumed a planetary importance, for exceeding the limits of a geographically or historically localized 'culture' or 'civilization' (Mehta 1976: 462-63).

(ii) Lynn White Jr. (1967), on the other hand, opines that it is the Judio-Christian concept of anthropocentrism that is responsible for ecological crisis. Judio-Christianity by establishing dualism between man and nature made it God's will that man exploit nature for his proper ends. While the first view ignores the significance of discontinuities within the West, White Jr.'s view loudly superimposes ecological crisis on Judio-Christianity, which is not true. Though it is true that God made nature for man to use, he according to Christian theology did not meant it for man's indulgence. Indulgence, not, use that causes ecological crisis. Franciscans, the ascetic school (Jains in Indian can be their equivalents) bears witness to the limited and minimum use of nature by man. Further, though man can use nature, there prevails the overall supervision of God who is the creator of both man and nature. Therefore, it is wrong to trace the roots of ecological crisis to Judio-Christianity.

(iii) Unlike Heidegger, who sees continuity from Plato to Nietzsche via Descartes, I would like to maintain that there is a significant discontinuity between the modern and the pre-modern theories. As mentioned already, in this paper, I would locate the rise of technology within the prisms of these discontinuities. It is Descartes who marks this discontinuity. It is not just the dualism between man and nature (I—It dichotomy) that is central to Descartes but the nature of the man, the Cartesian cogito, which accounts for man's domination on the nature. The implications of the Cartesian concept of man are far reaching.

This concept of man, well formulated in the Contract philosophies, the followers of Descartes, is a hypothetical construct. This construct is asocial, ahistorical, and implies 'amnesia' of the past. In other words, it is axiomatic and scientistic. These theorists bring this hypothetical man to the centre of social reality for the first time in human history. He is made the source and substance of social reality and its institutions. This is the basic assumption to the three Social Contract Philosophers, viz., Hobbes, Locke and Rousseau. The overt disagreements amongst the three are with regard to the attributes of the man-in-the-state-of-nature, i.e., whether he is egoist, selfish (Hobbes), noble savage (Rousseau) or rational and good (Locke). It is my contention here that this is the basic assumption, which not only marks the discontinuity in western thinking but also becomes the foundation of modern thinking. This assumption, along with its corollary, that is, society and its institutions created by this hypothetical man, what I shall call 'artificial society', underlies modern thought. This version of man has reference to the 'natural man'. It is interesting to note that the three Contract philosophers tried to provide empirical justification for their 'hypothetical' construct of man. Their inability to find an empirical referent for this idea, far from repudiating it, only served to reassert it as self-evidently true. The possible reason for the emergence of this hypothesis and the paradigm shift there after seems to be the developments in astronomy.

Thomas Kuhn, commenting on the potentialities of the Copernican revolution, observes, "though initiated as a narrowly technical, highly mathematical revision of classical astronomy ... [it] becomes one focus for the tremendous controversies in religion, in philosophy and in social theory, while during the two centuries following the discovery of America, set the tenor of modern mind" (1959: 2).

The Copernican revolution caused the shift from geocentrism to heliocentrism. The Sun was brought to the centre of the universe and earth the dwelling place for man was relegated to the periphery. Along with earth, man also seems to have lost his importance. This lost status of man seems to have been compensated by way of anthropocentrism—an existential need indeed!

The available versions of man in pre-modern theories or the natural man were not capable of discharging the new responsibilities once held by God. Hence, the need to postulate a bloated version of man. In this respect, though man replaces God, but in the manner of rendering his centrality with certainty, rationality, exclusiveness in the construction of social institutions, there is a remarkable continuity between modernity and tradition. This concept of man is all-powerful. He was decorated with concepts like freedom, liberty, individualism, etc. The implications of this new postulation, and its implicit discontinuity, are various. Some are stated below:

(a) Accepting the new assumptions meant a rejection of the 'neutral man' and his primary bonds (Eric Fromm) and the intermediary natural social institutions, which constituted his social life. These institutions, to name the prominent ones are family, language, market, culture, tradition, local authorities and customary law, etc., have been either rejected as false or transformed beyond recognition to suit the new theories. For instance, see what Rousseau has to say about the institution of family and how he renders it in the new form. He, while reducing the importance of family referring to even natural institutions in ancient societies, says:

> The most ancient of all societies, and the only one that is natural, is the family: and even so the children remain attached to the father only so long as they need him for their preservation. As soon as this need ceases, the natural bond is dissolved. The children, released from the obedience they owed to the father, and the father, released from the care he owed his children, return equally to independence. If they remain united, they continue so no longer naturally, but voluntarily; and the family itself is then maintained only by convention (1952: 4).

In contrast, see the fundamental status given to family in the classical philosophy. For Aristotle, family, which is sustained by hierarchy, is a basic social institution. This institution is not based on choice but is an extension of nature. Unlike modern philosophers, for him the whole precedes parts and anything like individual centredness is a myth or a fiction, as the birth of one already presupposes coming together of two. Therefore, for Aristotle, it is a basic institution. He says in *Politics*:

> First of all, there must necessarily be a union or pairing of those who cannot exist without one another. Male and female must unite for the reproduction of the species—not from deliberate intention, but from the natural impulse, which exists in animals generally as it

also exists in plants, to leave behind them something of the same nature as themselves (1986: 3).

Further, for him:

The first result of these two elementary associations [of male and female, and of master and slave] is the household or family. Hesiod spoke truly in the verse:

"First house, and wife, and ox to draw the plough."

The first form of association naturally instituted for the satisfaction of daily recurrent needs is thus the family (1986: 4).

Rousseau, unlike Aristotle, reduces the importance of family by treating it as mere contract, debunks its fundamental status by treating it as transitory and contingent, thus issuing out a radical discontinuity between the pre-modern and modern theories.*

(b) The implications of this discontinuity to theory and reality are enormous. The theories based on these assumptions (axioms) have become abstract, not the abstract of the pre-cognitively given but abstractions which is completely divorced from reality, having roots 'somewhere else'. While accepting that all theories involve abstractions, and real theory may be a misnomer, what is to be noted here is that these theories have completely fallen out of reality—an eternal damnation, a second fall.

On the other hand, the pre-modern theories are abstractions of the real, which is pre-cognitively given. This process of abstraction from the pre-cognitively given is described by 'transcendence' and the theories are often characterized as 'ideal'. These terms cannot be employed to the modern theories with the same import.

The presence of this gap seems to be haunting the postmodern era. Various postmodern attempts can be seen as forms of bridging this gap by resurrecting the natural features and, in doing so, seeking to reduce the abstractness of the assumptions. The contributions of Hegel, Darwin, Marx and Freud can be viewed as resurrecting the sense of history, evolution, social realism and pre-conscious, respectively. There

---

* However, it is another matter that like Aristotle, he too retains the patriarchic model by portraying the contractual relation between father and children. These implicit shades of patriarchy in the Enlightenment thinkers have been assiduously explicated by feminist, particularly S.M. Okin's analysis of Kant is noteworthy (1989). So there are two points to be noted one, Rousseau undermines the primacy of family, yet he retains the primacy of patriarchy.

are also the attempts to transform the given natural institutions to approximate to the theories and/or to create new social realities to provide content for the new theories.

Modern science and technology comes to serve this gap by transforming or creating realities. It is in this context we may have to locate the oft-made analysis that technology, though used by man, used him in turn. This is wrongly formulated, as technology is not so much used for man's ends but to serve the new concept of man, a hypothesis. It is this bloated concept of man, which is at the root of the modern crisis. It not only treats the nature as the 'other' but also the natural man, forcing the latter to be 'free', that is, freed from his natural settings and forced into these new hypothetical constructs.

My contention is that it is this discontinuity between the pre-modern and the modern theories, which is at the root of the technological culture causing ecological crisis. Consequently, though I would go along with Heidegger in accepting the acuteness of the crisis, I would not endorse his diagnosis. To my view, it is not correct to attribute the ecological problems or the rise of technological culture either to subjectivist-metaphysics or to the Judeo-Christian concept of anthropocentrism. Both these pre-modern schools though seemingly anthropocentric could not give rise to technological culture as theories in pre-modern schools are of a speculative character, in that they assume the given real and attempt to arrive at the teleology of this real through their theories. Here, the purpose is transcendental than empirical. This transcendental aspect of pre-modern theories could not have made empirical developments such as technology possible.

The pre-modern times have not witnessed much technology and it is a modern phenomenon and needs to be distinguished from craft. The latter is not backed by theory whereas technology is backed by well-articulated scientific theories—which make the duplication possible. Crafts are rooted in the day-to-day experience of a people, having explanation as post-facto. Modern science is rooted in theory; which has to be applied to reality by technology either to transform or to erect new realities. Modern science insofar as it is rooted in this empiricist attitude is bound to be technological. Scientist's ignorance of the technological potentialities of a scientific theory is no justification to claim the purity of modern science. The pre-modern sciences, on the other hand, like pre-modern theories, remained at the level of ideal; they were not intended for use purposes, except in warfare and medicine. As already stated, this is not to suggest that pre-modern theories were exclusively ideal. For their ideal is the ideal of the ideal of the pre-cognitively given.

The purpose is more to understand nature, to transcend it, rather than transforming it. In other words, their teleology is transcendental and not technological.

Further, the concept of man in the modern theories is different from the man of the pre-modern theories. The concept of man in the latter is the extended version of the naturally given man, a state to be acquired by man through constant practice. Such a man is not disposed to creating large-scale eco-technological crisis. Thus, the rise to technology and subsequently ecological problems are to be located in the discontinuity brought of by cartesianism. Rather than in the continuities of western metaphysics as held by Heidegger who fails to recognize them.

Acknowledgements

I wish to thank Sasheej Hegde and A.V. Afonso for suggesting me important corrections in the paper. The earlier version of the paper was read in an ICPR sponsored symposium on *Technology, Ecology and Heideggerian Ethos* at Goa University during 1-2 December 1988.

References

Aristotle (1086), *Politics of Aristotle*, translated with an Introduction by Ernest Barker, Delhi: Oxford University Press.

Durbin, Paul T. (ed.) (1979), *Research in Philosophy and Technology*, Vol. 2, Connecticut: Jai Press, Inc.

Heidegger, Martin (1962), *Being and Time*, New York: Harper and Row.

—(1968), *What is called Thinking?*, New York: Harper and Row.

—(1977), *The Question Concerning Technology and other Essays*, New York: Harper and Row.

—(1981), *Nietzsche*, vol. 1. London: Routledge and Kegan Paul.

—(1982) *On the way to Language*, New York: Random House.

Kuhn, Thomas (1959), *The Copernican Revolution*, New York: Random House.

Mehta, J.L. (1976), *Martin Heidegger: The Way and the Vision*, Honolulu: The University Press of Hawaii.

Okin, Susan Moller, "Reason and Feeling in Thinking about Justice," *Ethics*, 99, No. 2.: 229-49.

Rousseau. J.J. (1952), *Social Contract and Discourses*, London: J.M. Dent & Sons Ltd.White, Lynn, Jr. (1967), "The Historical Roots of our Ecological Crisis," *Science*, 155: 3767.

8

Anthropocentric Teleological Eco-ethics

R.C. Sinha

The present paper concerns new area and new direction in applied ethics. The purpose is to give philosophical perspective and illustrate the connection between ethical principles and ecology. It also aims to dispense with the criticism that modern ethical philosophy is not concerned with the problems of eco-ethics and too much preoccupied with linguistic issues of the meaning of ethical concepts. Here, I have made the most modest attempt to discuss ecological problems and demonstrate as to how philosophical perspective can be relevant to the solution of environmental problems we face in our day-to-day life.

The title of this paper is quite suggestive. I have tried to pinpoint that eco-ethics is both anthropocentric and teleological. My stand is that eco-ethics, independent of man, is not possible. Man is the sole bearer of values. Man as a valuer is essential. Ecology serves human purpose.

Ordinarily, the problem of ecology is dealt from natural and scientific perspective and not from philosophical perspective. Recently, philosophers have started taking interest in ecological issues in a systematic philosophical manner. The ethical standards are applied to ecology. This gives birth to eco-ethics. At the present time, there is a widespread concern over moral issues in ecology. Eco-ethics is environmental ethics. The environment and ecology cannot be treated as synonymous. They cannot be interchangeable. I have preferred the expression 'eco-ethics' instead of 'environmental ethics'.

In postmodern era, philosophic ethics has been faced with relatively new challenges. This challenge has been posed from many quarters from outside as well as within philosophy. It is heard in the form of calls from diverse environmentalist and policy making organizations for the study and teaching of ethics relating to environment. Philosophers started reflecting on the growing awareness of the many moral, social and political issues, which beset the ecological problems. In opposition to traditional beliefs, a new morality seems to have emerged, as evidenced from the changing attitude towards environment. New problems have also been created by scientific and technological changes. Accordingly, many issues of eco-ethics and bio-ethics, which focus on matters of nature and life, have cropped up. New philosophical concerns have been evoked especially by a growing number of moral controversies relating to ecology. In the past also, there seems to have been more agreement about matters of right and wrong conduct relating to environment. The classical philosophers shared many beliefs about the rightness of wrongness of particular acts. They were more concerned about the appropriateness or inappropriateness of human behaviour towards nature. Philosophers concentrated on finding principles, which would support shared moral convictions and serve to resolve environmental problems. Today, by contrast, there is relatively more controversies on moral issues relating to environment.

In the twentieth century, the work of philosophers has focused more on problems of theory construction than on problems of application. This focus has been evident in the seemingly inordinate amount of attention given to meta-ethics by analytic philosophers. Analytic philosophers speculate over the nature, scope and even the possibility of moral reasoning. They do care little on substantive issues of environment. In fact, most analytic philosophers have felt that it is none of the business of philosophers to address issues relating to ecology. Despite the strong movement of linguistic analysis, there are contemporary ethical philosophers who have proposed substantive ethical theories, and some have even questioned the distinction between meta-ethics and normative ethics. Those influenced by Kant have argued that moral rules can be established on the basis of universal applicability and respect for persons. Contemporary utilitarians argue that rightness or wrongness of action is determined by the values of consequence. The skepticism over the validity of ethical reasoning and opposition between various schools of ethical theory have created the impression that philosophy has little to offer in the way of solutions to everyday, practical and environmental concerns.

Since late 1960s and the early 1970s, philosophers have been addressing contemporary moral issues and whole new fields of eco-ethics have developed. In the past ten years, there has been a vast increase in the number of conferences, societies, journals and texts devoted to the subject of environmental ethics. At the same time, on the philosophical level, new models have been developed to provide procedures for making practical decisions and resolving disputes relating to ecological problems.

In many respects, the eco-ethics has shown that philosophers do have something to offer by way of clarifying issues and positions, and even by showing how, or to what extent, one or another theory can be applied to ecology. Nevertheless, many critics feel that there is still a serious gap between ethics and eco-ethics. Ethics is theoretical where as eco-ethics is applied one. One reason for the criticism of ethics is that much of the work in applied ethics presupposes the position of one or another school of philosophy and hence does not face up to the problem of opposing philosophical views. In other case, it turns out that one or another existing theory is simply not refined enough to yield answers, even according to the principles it lays down. In still other cases, applied philosophy is done without any attention to theory at all.

It is, therefore, important to investigate the relation of ethics to eco-ethics not only from the point of view what philosophers have to offer to the solution of ecological problems, but also from the point of view of seeing how philosophical ethics itself might be improved by considering problems of application in the field of ecology. How does the question of the applicability of ethical theory bear upon the question of ecological facts? In addition to the test of internal coherence, should there also be a test of completeness, based upon a theory's ability to resolve practical environmental disputes? How can ethical principles be elicited from the subject matters in which moral questions arise in order to resolve differences between theories or to construct a more comprehensive ethical view?

I would like to make it clear that moral philosophizing and the application of moral standards in ecology is not limited to moral philosophers. Practically, everyone philosophizes about values. A person other than philosopher also uses moral theories. Philosophy, however, is the only discipline engaged in the study of ethical theory as one of its special subject matters. Philosophers attempt to philosophize and justify the theories they propose. When one seeks answer, not only to what is right or wrong in particular cases, but also to questions of correct principles, it seems only natural to turn to philosophy for answers.

The problems of philosophizing arises quite naturally when anyone begins to reflect upon his or her moral practices, questioning the justification of actions or reasons for judgements. Disagreement is often the source of philosophizing. It also arises when people are genuinely perplexed about what they should do. In their attempts to resolve ecological controversy or remove doubt about environmental problems, moral philosophers seek reasons for or against particular actions. In so doing, philosophers often appear to personal codes or socially accepted rules. The moral standards appear to be conflicting. There is wide range of disagreement among philosophers. The applicability of the standard of moral judgements is not uniform. An accepted rule may also seem applicable to a particular case. Philosophers proposed principles which will justify the rules, resolve conflicts between them, determine the range of their application, justify exceptions or clarify their meanings. Moral reasoning does not always proceed in this way, of course, for persons may begin by appealing directly to accepted moral principles or to contextual values, and some simply refuse to reason in support of their opinions. Traditionally, it has been the philosopher's job to sort out such responses.

The expression on values like efficiency, worth, utility, significance, speed are very much current in postmodern world. The modern world was essentially rationalistic and materialistic. Its emphasis was on centrality of problem. Postmodernity upholds multi-narratives. Postmodernity explodes the myth of modernity. But, the postmodern world is pluralistic and relativistic. Modernity is the age of machine. Postmodern age is the age of technology. The values of life have also changed in postmodern world. The modern age was the age of development. The reckless use of resources was the hallmark of industrialization of modern age. The postmodern age gives attention to ecology and its value for human existence. In postmodern scenario, the expression like 'utility' or 'worth' has been used in moral discussion. These terms indicate the fact that possession of values is a pervasive phenomenon. Man endeavours to analyze values. Man tried to choose between alternative values of life. Man deliberated upon values of life to plan the course of actions. Man also passes judgements about the motives and actions of human beings. Making judgements or evaluation of human conduct is a part and parcel of day-to-day business of life.

This analysis raises two questions: first, how one 'actually' conducts himself in relationship with others and secondly, how one ought' to conduct in an interpersonal relationship. In the former case, we describe behaviour and in the latter, we 'evaluate' actions and motives as right or wrong. This distinction between 'actual' and 'normative' pinpoints the distinction between positive science and normative science. The positive science is concerned with the 'is' aspect of a thing whereas normative science is concerned with 'ought' aspect of a thing. The scientific statements are descriptive whereas ethical statements are evaluative. The distinction between 'actual' and 'normative' points to a separation of everyday morals from ethics. Ethics is an axiological theory. Axiology is the science of values. Ethics is a valuational theory. It implies a reflective analysis of value preferences, behavioural norms and codes of conduct in a specific spatio-temporal context. Accordingly, ethics is normative. It deals with norms or customs of society. It is concerned with standards of human conduct and not merely with actual behaviour of human beings. However, ethics is a normative study of human conduct in a particular society over a period of time.

Environmental ethics mark a significant departure and signifies an all-pervasive whole wherein subsists all natural beings. In other words, environment includes both biotic and a biotic species. Ecology implies an interrelationship of all these species and their environment. Eco-ethics leads to balance of ecology and considers destruction of natural resources as unethical. The moral sense guarantees sustenance of human existence through balance of ecosystem. Sustenance means managing, conserving and preserving natural resources and all living beings. Ecological sustenance means a sort of fellowship between humans and non-humans.

The champions of environmental ethics have pointed out to the dichotomy existing in ethical theories regarding man-nature relationship. The environmental degradation in terms of exhaustion of natural resources started long ago. But, the recognition of this degeneration and, therefore, a response to it started only about two decades back. In 1969, the Secretary General of the United Nations warned that inhabitants of the world have perhaps ten years left to improve the human environment. If environmental problems are not addressed soon, he said, then they will have reached such staggering proportions that they will be beyond the capacity to control.

Now let us make a distinction between anthropocentric ethics and eco-centric ethics. Antrhopocentric ethics is man-centric whereas eco-ethics is nature-centric. The trend and tendency of twentieth century ethical thought is primarily man-centric. The human beings are special bearers of values. Ecology being physical has been kept outside the purview of ethical discourse. A clear-cut distinction between moral and non-moral actions has been made. Ethics is neither concerned with the activities of trees, plants and nature nor with activities of beings other than human beings. We cannot pass a moral judgement on the activities of non-living beings or animals. I think that eco-ethics will be meaningful if it is anthropocentric one. I do not subscribe to the distinction between anthropocentric ethics and eco-centric ethics. On the contrary, I contend that eco-ethics should be anthropocentric. By the expression anthropocentric, I mean that ethics is meaningful with reference to man. Eco-ethics is man-centric because man's attitudes and values are linked with ecology. There is no watertight compartment between anthropocentric and nature-centric ethics. Man is the moral agent. Ecological behaviour in itself is neither moral nor immoral. It assumes moral significance when linked with human affairs.

To pass a judgement is to use a moral standard. We agree that standards of morality conflict with each other. Moral philosophers have offered a variety of alternative standards. In the history of ethical thought, there are two sorts of theories. One is teleological and the other is deontological. Teleological is consequentialist whereas deontological is intrinsic. According to teleological theory, the worth of an action depends upon its consequences. Teleological theory originates from the Greek word 'telos'. Telos means purpose. It suggests an action is right if it leads to good consequence and an action is wrong, if it leads to bad effects. Action is evaluated as right or wrong if it serves human purpose. Purpose is termed as good or bad depending on benefits or quality of consequences that it brings about. Teleological theories have either propagated promotion of self-interest or have advocated the greatest good of the greatest member. However, over the period of time, the teleological position has changed from merely counting the quantity of consequences. Hence, in recent times, it has taken the shape of 'ideal utilitarianism'. The utilitarianism has often been associated with teleological theories, since the purpose is often weighed in terms of benefits or utility. Nevertheless, the 'rule' or 'ideal utilitarian' like G.E. Moore have asserted that the values cannot be defined in terms of interest nor can they be equated with pleasure. "A right action is one that brings into existence the maximum amount of intrinsic value of quality taking all the consequence into considerations."[1]

Deontological theories, on the other hand, uphold that the rightness and wrongness are intrinsic properties of some actions or types of actions. Actions are immediately right or wrong regardless of their consequential merits. Deontological theory stresses our obligation to do our duty without weighing its consequences. The three staunch supporters of the kind of moral trends have been Joseph Butler, W.D. Ross and Immanuel Kant. Butler says that we discern immediately where our duty lies and accordingly the basic principles of morality are justice, honesty and truthfulness. According to Ross, we immediately know that certain types of action are always right or wrong. For example, we "ought not to speak lie" is directly known to be morally binding. Kant advocated categorical imperative, which are unconditional laws.

The twentieth century moral philosophers asserted human-centric ethics by stating a case for moral relativism as opposed to categorical imperative of Kant. Taking an example from anthropological evidence, it has been argued by cultural relativists that there is an indefinite variety of moral standards. So, there is no absolute standard of morality. Morality is a product of certain historical contexts and certain specific customs. There have also been other approaches to ethical theory which questioned the very rationality or reasonable justifications of standardized behaviour. But the end of twentieth century is definitely marked with a shift in the tendency. In certain areas, a relativistic ethics needs to be reviewed. This brings us to the analysis of application of standard ethical theories on man-nature relationship.

The anthropocentric teleological ethical theory translated into environmental ethical context would imply that ecology should be preserved because it is useful for human existence. The abiotic entities like rivers, forests, mountains, hills, etc., are subservient to human purpose besides being providers of basic needs and vital breath. Therefore, green fields and forests should be preserved because they provide us with fodder, medicine, fuel, etc. The hills and mountains, sunshine and sunset of Kanyakumari, Palm trees of Kovalam beach of Trivandrum, the flow of pure water in Ganga at Haridwar and Varanasi provide us with an aesthetic delight. It kindles the pious feeling of sacredness. The natural resources, both biotic and abiotic, are essential for life. There is a great controversy over major developmental projects in India, such as the Sardar Sarovar dam on the river Narmada, the Chilka lake in Orissa, the Konkan Railways, etc. Actually, objections to these projects pertain to the extent of environmental destruction and uprooting human settlements. But, some protagonists in the name of development justify these projects. In the postmodern world, sustainable development has become the new buzzword. Sustainable development is development that meets the needs of the present without compromising the ability of the future generations to meet their own needs. This view has been offered by the World Commission of Environment and Development in its report "Our Common Future". Economists define it as an economic progress in which the quantity and quality of our stocks of natural resources and the integrity of ecosystem are sustained and passed on, unimpaired, to future generation. The reason for destruction of environment in the name of development lies in mistaken separation of man from nature as Arundhati Roy has put it very rightly that "nature needs to be saved from the humans for the humans."[2]

The one-sided approach of seeking only human interest has backfired on man himself in a subtle way man uses nature for his benefits. Men have been crude enough to exploit other beings rudely; for instance, the experiments conducted on animals for the medicinal and cosmetic benefits of human beings, clearly point out to the arrogance of man. Man endowed with reason and technology considers himself different and superior to other biotic and non-biotic things. It seems that there is a logical entitlement to some humans to treat other beings with disrespect and as being grossly unequal to them. In other words, nature becomes a resource and man the consumer.

The ecosystematic ethics would take a position that nature or everything other than humans is to be respected because it is our duty to do so. Deontological theories have always made duty or commitment the basis of human conduct because the moral object, i.e., the issue in question is inherently moral. So translated in deontological ethics, this position would demand friendliness and fellowship with nature because it is by itself worthy and therefore, it is the duty of human beings to protect and preserve nature. Life in general is seen as having a value. Life ought to be respected. Aldo Leopold, whose works can definitely be referred to an environmental treatise, once stated that "a thing is right when it tends to promote the integrity, stability and beauty of the biotic community. It is wrong when it tends to be otherwise."[3]

Hence, the views of adherents of eco-ethics are that human beings are not at the centre stage of ethics. Human beings are members amongst other members of the biotic community. In one sense, human beings endowed with rationality are different from other members but certainly not unequal. This position clearly makes the whole ecosystem, "morally considerable". There is a pressing necessity to find an amicable solution to sort out the man-nature divide. What we need to posit as new norm is an 'eco-ethics', where there is a fellowship between man and nature, and are yet equally entitled to moral approvals. To my mind, anthropocentric teleological eco-ethics needs to be purposive and futuristic in its approach since the friendship between man and nature needs to be everlasting. The basic need of fresh air, water and food would remain and this is what entitles nature the commitment from humans to preserve it and sustain it. To generate friendliness between man and nature, it is important to realize that this can be done only through the right set of values. This would include dispositions and conducts of natural respect and amicability. Anthropocentric teleological eco-ethics include wider range of biotic and abiotic things in its purview. Further, it needs to be extended to being 'universalistic' since relativistic ethics has already led to partial and incomplete set of norms. Anthropocentric teleological eco-ethics demands a global network since environment is all-pervasive. It needs to be recognized at global level. We are not evaluating governments and societies in a parochial context while trying to form anthropocentric eco-ethics.

The new eco-ethics give due importance to all biotic and abiotic entities. Some of the critics may say that all living beings cannot be treated equally. The difference cannot be overlooked. In environment, there are species, which are required to be preserved. The principle of equality may be in accordance with Rawlsian "difference principle"[4]. Rawls' difference principle contemplates that advantage to one being should not cause disproportionate disadvantage to other. Thus, alliances between man and nature ought to be based on the inclusion of wide range of interest—present and future, human and non-human. We try to formulate this theory and translate it into practice. We are already on our way to a commitment to nature, to our present and future generations and to a sustainable man-nature relationship.

From the above critical analysis, it is clear that the reason for eco-ethics is teleological rather than deontological. We preserve ecology and maintain balance because it is beneficial for human beings. Some of the critics may point out that development needs utilization of natural resources. The objections and protests by Narmada Bachao activists and other environmentalist may hamper the development. Development at the cost of wanton destruction will herald doom to future generation. So, the sustainable development and use of natural resources will be appreciated. The eco-ethics believes that sustainable development should not be hampered but it should not upset ecological balance. Ecological imbalance will predict doom for new generation. So, we have to preserve ecology because it serves our purpose of life. In postmodern world, it is not possible to uphold deontological theory, which contends that we should respect ecology since it is our duty to respect it.

The eco-centric ethics put a challenge to anthropocentric ethics. Eco-centric ethics has changed the whole way we have thought about right and wrong in classical ethics. E.F. Schumacher is correct in arguing that our traditional economic institutions have confused us about ethics. Ethics, especially Christian ethics, condone mores or moral customs that constitute "an act of violence against nature which must almost inevitably lead to violence between men."[5] Hindu ethics believe in love for all living beings whereas Christian ethics advocate love for human beings. Christianity thinks that lower beings should support the higher beings. Lynn White observes that "Christian traditions have erred in destroying animism and in supposedly sanctioning an absolute human right to dominate nature."[6] Schumacher and White have concluded that traditional ethical theories do not take adequate account of planetary well-being. The environmental ethics is required to provide a theoretical basis for taking into consideration the needs and interest of nature, rather than merely the needs and interest of humans. W.H. Ferry believes that we need a new ethics to deal with the environment. Ferry, in his article entitled, "Must we rewrite the constitution to control technology?", in *Technology and Society*[7] upholds this view. Likewise, biologist Wayne Davis has argued that "the time has come where we must all develop respect for the air, water, land and the living things. We are all dependent upon the same life support system of earth and must protect it if we are to survive. I hold these truths to be self-evident. All living things are created equal and are interdependent upon one another."[8] Pursuing a slightly different line, attorney Christopher Stone has argued for the recognition of "the legal rights of natural objects on the grounds that in principle legal rights have been and can be expanded to include 'inanimate things.'"[9] A historian at the university of California has become well known for his belief that "rocks have rights". Australian philosopher Peter Singer has become the recognized international authority on the question of animal rights. Singer has argued, for an end to 'speciesism' and for recognition of animal rights on the grounds that, since animals suffer, there can be no moral justification for refusing to take that suffering into account in our ascription of rights."[10] This is quite true to say that humans have ignored the well-being of ecosystem.

Generally, when philosophers propose adoption of a new environmentally-oriented ethics, they are depending on one of two possible approaches, viz., an expanded rights views and an integral view based on ecosystematic considerations. Expanded rights view holds that right should not be limited to human beings, rather it should be expanded to other living beings. Proponents of the expanded rights view adhere that "rocks have rights". They see new environmental ethics as another "liberation movement", but for flora, fauna and rivers instead of for blacks and women only". Similarly, the advocates of the ecosystemic view believe that what makes our actions morally right or good, is the way they affect ecosystems. Although ethicists have great empathy for proposals for a new environmental ethics, especially one based on ecosystematic considerations, both types of proposals seem to me to suffer from fatal conceptual flaws. The expanded rights view of environmental ethics holds that rocks and rivers, as well as sentient beings, are said to have rights. If such an expanded rights view proposes merely that the class of right holders be enlarged to include all beings that can feel pleasure, pain or joy. Then, of course, there is nothing problematic about accepting such a view. Bentham has spelt out his utilitarianism long ago, stipulating that each sentient being, not merely each person or human, was "to count for one and more for more than one". His reasoning was that insofar as a being could suffer, it was worthy of consideration—a point of view accepted by a number of moral philosophers as well as by most contemporary thinkers. People generally agree that it is wrong to mistreat animals, not because they are moral persons but because they suffer, and needless suffering is to be avoided.

Insofar as the expanded rights position calls for including beings that are not sentient in the class of right holders, then it is both new and problematic. It is unclear why one should respect a thing, for example, a rock or river, if it experiences no pain, elation or suffering. What is it non-sentient things that deserve our respect? Why, apart from anthropocentric considerations, should poison oak have right? Why, other than for anthropocentric reasons, should one not dump toxic wastes in place where they will pollute groundwater? Of course, there are sound anthropocentric teleological reasons, why persons should not destroy rare plants or despoil landscapes? My reason for that belief, however, is not that rocks or rivers have rights but that needless destruction bespeaks human callousness, myopia, and perhaps greed—all of which are wrong on anthropocentric grounds. The point of view of the expanded rights theory appears to be that "there is some other, inherent, non-anthropocentric reason for not picking flowers or destroying rocks. But, such a notion of inherent value is puzzling. What would it mean to say that everything has rights.[11] How then, would any action be possible that did not harm something? If everything had equal status, then there is no need of controversy over rights. If we take the position that everything has equal rights, the humans would be prohibited from killing deadly bacteria infecting human life. The concept of 'Ahimsa' of Jainism conceives the extreme view of non-injury to bacteria floating in the air. What could a right mean in such a situation? I think, these will be conceptually wrong to hold such a position. The inherent value to all sorts of things like rocks, and at the same time all beings that there was nothing special about being alive or sentient or intelligent?

Yet, another problem with the expanded right position, insofar as it calls for recognition of the alleged rights of non-sentient beings, is that it fails to take into account that all values are necessarily values for us. There is no such thing as a value in and of itself, without a valuer. If non-sentient beings cannot have values, what would it mean to arrest their values? How can one visit good or evil on something that has no thoughts or feelings, even in a potential sense. The point of the proponents of the expanded rights view, however, seems to be that they can provide a rationalistic, analytic, naturalistic argument for their appeal to expand the class of right-holders, and that they do not see their position as mystical or theological.

The integral eco-ethics conceive that man is the integral part of ecosystem. There is no division between man and nature. According to the exponents of integral environmental ethics, recent industrial history is a biotic haemorrhage, to be remedied by judging all human actions in terms of their effects on all the ecosystems of the planet. According to this view, whatever promotes ecosystematic well-being is said to be good and right, and whatever jeopardizes it to be bad and wrong. A basic problem arises, however, why consider a whole that is not a conscious sentient being? What reasons are there, apart from human well-being, for ascribing moral worth to some ecosystematic whole? In response to such a question, a supporter of ecosystematic ethics is likely to respond that the earth has a beauty, a balance of nature that is inherent in it. But, the problem with which a response is that these are values that the earth and its ecosystems have only as objects of considerations by conscious, sentient human beings. It is not clear that the earth has these alleged properties apart from some viewer or valuer. Hence, it is not clear that there is some non-anthropocentric basis for affirming on ecosystematic ethics. Moreover, even if there were inherent values in ecosystem, apart from human or divine ascription, it is not clear what that might be. Presumably, the purpose of affirming an ecosystematic ethics would be to provide a value system according to which actions could be judged right or good, if they promote the well-being of the plant. But what might it mean to help or hinder the well-being of the biosphere or to maintain some balance of nature? Ecosystems regularly change, and they regularly eliminate species. After all, the dinosaurs are gone. It is unclear how contemporary proponents of ecosystematic ethics can assert some non-anthropocentric theoretical justification for their repeated claims that humans should not wipe out species, when nature does this herself. Of course, for clearly anthropocentric reasons, I do not believe species should be destroyed by humans. But it is a different matter to claim that species should not be destroyed for their own sake. Why is the elimination of species by humans different from their eliminations by natural selection? How can it be wrong for humans to do what nature does if the only grounds for alleging wrongness are theories about maintaining a balance of nature? It is generally maintained that the balance of nature is not a static thing. This being so, it is unclear how adherents of ecosystematic ethics can maintain that certain ecological changes are good whereas others are not. The differences cannot be merely that what happens naturally is good and that what happens through human interference is bad. This will give us sanction to a purely stipulative definition of ecological and moral goodness and badness.

The main difficulty besetting ecosystematic ethics is that is somehow derives an ethical 'ought' from a scientific or ecological 'is'. In other words, some natural characteristics of ecosystems are taken as normative, as things that should be preserved. Some moral philosophers will say that one cannot legitimately deduce some moral duty from an existing state of affairs. The mere existence of thing is not a justification for what should be. Thus, the inference of ecosystematic ethicists from 'is' to 'ought' must be mediated by value judgements.

The trends of anthropocentric teleological ethics are found in Hinduism, Buddhism and Jainism. The Buddhist ethics is deeply concerned with anthropocentric eco-ethics. The Buddhists monks always adhered to intimate relationship between living and non-living beings. The Buddhist ethics comprehends biotic and abiotic within its purview. The Hindu ethics is teleological because nature serves life. The Buddhist concepts of 'Karuna' and 'Maitri' are quite significant. Some contemporary Hindu thinkers like Sri Aurobindo and Rabindranath Tagore also uphold that ecosystem must be valued. Sri Aurobindo says that to attain bliss human beings see the divine everywhere. Finding bliss in life, however, requires an effort to unite with the divine at every level of existence. Sri Aurobindo's contentions is theo-centric eco-ethics. He says that one must remember that will to live draws its sustenance from the joy of life available in all animate and inanimate objects in nature. I do not subscribe to theo-centric eco-ethics because atheist and Marxist can question the existence of God. But neither theist nor atheist will question the existence of man and his goal in life. In sum and substance, the preservation of ecology is the prime concern for human well-being. We can say that the task of improving the environment is an urgent one. The solution of ecological problems cannot be short term. It requires long-term changes in perception. According to Indian thought, the moral philosophy plays a major role. This is because ethics helps to determine the decisions we make. The way Hindus answer, their ethical questions affect the way they live their lives.

Notes

1. G.E. Moore, *Ethics* (see in particular chapter "The Objectivity of Moral Judgment"). Oxford University Press, 1971 (reprint).

2. Arundhati Roy, *The Greater Common Good*, NBA Trust, Indian Book Distribution, 1919, p. 12.

3. Aldo, Leopold, "Land Ethics" in Robin Attfield and Andrew Belsey (eds.), *Philosophy and Natural Environment*, Cambridge University Press, 1944.

4. John Rawls, *A Theory of Justice*, Oxford University Press, 1973, p. 60.

5. E.F. Schemacher, *Small is Beautiful*, New York, Harper and Row, 1973, p. 57.

6. Lynn White, "The Historical Roots of Our Ecological Crisis", *Science*, 155 (10 March 1967).

7. W.H. Ferry, *Technology and Society*, London, Adolison Wesley, 1972, pp. 158-159.

8. W.H. Davis, "The Land Must Live", *Environmental Law Review*, 3 (15 July 1972), pp. 5-7.

9. C.D. Stone, *Should Trees Have Standing?* Los Altos, William Kaufman, 1974.

10. Peter Singer, *Animal Liberation*, New York, Review of Books, Section 20 (April, 1973).

11. Holmes Rolston, "Is There an Ecological Ethics?", Ethics, 85:2 (January 1975), p. 101.

9

Environmental Ethics: To What Extent Does It Go Beyond Human-Centred Ethics?

Dinesh Chandra Srivastava

It has been fervently debated in contemporary philosophy as to which beings have and which do not have a claim to be considered morally important in their own right. An ethical theory is anthropocentric or human-centred if it denies such importance, independent moral status—to anything but human beings. Some philosophers have castigated human-centred ethics, as 'human chauvinism' on the ground that is cannot account for our moral relations with the natural environment. They have argued for a new ethic—environmental ethic—that extends the range of morally considerable cases beyond human beings. Non-human beings, whether animals and plants and other living beings or even non-living natural systems, are all credited with moral worthiness and hence independent moral status.

While the question of relationship between morally aware subjects and the objects in their environment is a central concern of environmental ethics, there is the problematic philosophical issue of determining *which* non-human really comes within the scope of intrinsic ethical worth. We can conceive of two extremes with narrow and wide scope. The narrow-scope view is that things in the natural environmental are morally worthy so long as they relate to human interest. In the wide-scope view it is the entire ecosystem that has moral worth independently of any relation to human interest.

The question the paper focuses upon relates to the wide-scope view. It is the question of whether radical environmental ethics can defend its radical stance with regard to the extension of independent moral status to almost the whole of the natural environment, or the entire ecosystem of the universe. Can moral value really reside in the ecosystem itself, which consists, in addition to living beings, non-living things such as mountains, rocks, rivers and sky? Can such natural systems and their independent elements constitute a domain of moral transaction, even though there appears to be nothing in such systems analogous to moral transaction in the human social domain?

It is interesting to witness the recent transition, in the West, from the anthropocentric liberal moral tradition to a new moral tradition of 'deep ecology' (owing to be Norwegian philosopher Arne Naess) that strikes us as essentially cosmo-centric in spirit. Deep ecology advocates our essential oneness with nature, where to be one with nature means to become part of "great Self"—where the great Self (with capital 'S') symbolizes what Naess calls "the living earth". This is evidently a cosmo-centric vision of our place in reality as espoused by traditional Indian philosophy. It may thus be wondered whether contemporary western ethics, in the guise of (wide-scope) environmental ethics, is unconsciously coming to terms with the spirit of ancient Indian philosophy.

The paper discusses the following three major problems:

1. The question of the extension of the ethical domain from the human to the non-human, and examination of the legitimacy of recognizing the ethical significance of non-human world, in all its variety, in its own right.
2. Reflection on the nature of ethics in the light of radical challenge posed by environmental ethics of wide scope.
3. Exploration of the possibility of the basis for a sound environmental ethic in Indian philosophy, especially in the light of the idea of deep ecology in the West.

Let me firstly discusses the narrow-scope view that things in the natural environmental are morally worthy so long as they relate to human interest. Some thinkers like B. Norton are of the view that environmental policies should be evaluated solely on the basis of how they affect humans. This entails a human-centred environmental ethics.

Although the classical utilitarians include animal suffering in their ethical calculation, a variant of utilitarianism, which enjoins us to maximize the surplus of human happiness over human unhappiness is one example of a human-centred ethic.

A human-centred theory of environmental ethics holds that our moral duties with respect to the natural world are all ultimately desired from the duties we owe to one another as human beings. It is because we should respect the human rights of everyone or should protect and promote the well-being of humans. We must place certain constraints on our treatment of the earth's natural environment and its non-human inhabitants.

The theory of human ethics has three main components. First, there is the acceptance of a *belief system* within which each moral agent conceives of others in a certain way. One's fellow humans are perceived as persons like oneself. Each is understood to be a subjective centre of conscious existence with the capacity to choose its own value system and live a self-directed life. Thus, each is seen to exemplify the same traits of personhood found in oneself. The second component is the *attitude* of respect of personhood found in oneself. This attitude is both a moral one and an ultimate one. It is moral attitude because it is universalizable. As a moral attitude, it is considered binding upon everyone alike. For the one who has the attitude of respect, each person is deemed worthy of everyone's respect, not just of one's own respect. This attitude is an ultimate one because its justifiability does not consist in being derivable from any more general or more fundamental attitude. The third component is a system of *rules and standards* considered valid in the domain of human ethics.

The above three components, viz., the belief system, the ultimate moral attitude and the set of rules and standards, are related to each other. When a moral agent accepts the belief system and conceives of others as persons, he or she takes up a certain outlook on the social world. Others are seen as belonging to a community of which one is also a member on equal terms with them. The community itself is so ordered as to make it possible for all individuals to live self-directed lives according to value system of their own liking, subject only to those constraints needed to give each an equal chance.

The above human-centred ethics is often called anthropocentrism. There are some environmental philosophers in the contemporary West (e.g., Richard Routleys) who prefer to call the above view with expression 'human chauvinism'. Routleys quotes, as a representative expression of 'human chauvinism', and 'a core principle of western ethics', the following:

> The liberal philosophy of the western world holds that one should be able to do what he wishes, providing (i) that he does not harm others, and (ii) that he is not likely to harm himself irreparably.[1]

Routleys concede that this principle imposes real restrictions on the freedom of human beings to deal with the environment as they desire. This principle entails a denial of independent moral status to non-human beings.

With the above cutting remarks of Routleys, let me now move to the wide-scope view which accords moral worth to the entire ecosystem, independently of any relation to human interests. We may call this view a radical environmental ethics, which can defend its radical instance with regard to the extension of independent moral status to almost the whole of the natural environment—the entire ecosystem of the universe. The philosophers and activists within the environmental movement have divided over the question whether our responsibilities towards our environments are ultimately responsibilities to human beings or to non-human entities directly. One important reason for this is that there is a division over what entities have independent moral status.

We will now look at this division within the environmental movement. In 1973, one of the founding fathers of environmental philosophy, the Norwegian philosopher Arne Naess, published a short paper entitled *The Shallow and Deep Long-Range Ecology Movement.*[2] As a result of Naess' influence, the terms 'shallow' and 'deep' have become common currency amongst environmentalists. A shallow position is one that reduces concern for the environment to concern for the interests of human beings. A deep view is one that is committed to "the equal right to live and blossom" of all forms of life. Alternative labels that are sometimes used are 'reform' and 'radical' environmentalism. But, another alternative ready to hand is the term 'green', which allows us to speak of *light green* and *deep green* variants.

First, the shallow, the reform or light green approach, as a practical approach to environmental problems, can be summarized in four propositions as follows:

1. Environmental problems are identified as changes in some environment, whether brought about by human action or not, that pose a danger to human health, comfort or even survival—in short to human well-being.

2. Where these changes can be controlled or reversed, any human being has reason to act in ways that will help avert the threatened harm; these will include changes in personal behaviour and supporting collectively adopted measures.

3. Human well-being is intimately tied up with the well-being of many other individual things, either because they are useful to us or

because we care about them directly; consequently, they are protected by actions designed to secure human well-being.

4. Since only human beings have independent moral status, the protection afforded to non-human as a consequence of their mattering to humans is all the protection it makes sense to demand.

The above four propositions define a position at one extreme end of the light green-deep green range. Its most fundamental feature is its restriction of independent moral status to human beings. At the opposite extreme is a position which accords independent moral status to all living things in general, or even to collective entities such as species, forests and rivers. This deepest position may be characterized in three propositions:

1. Problems are identified as changes in some environment that pose danger to anything within this unrestricted field.

2. Action to prevent these dangers requires that human agents recognized duties much more extensive than those recognized by traditional moralities which accord independent moral status only to human beings.

    Exponents of such deep green position often maintain that traditional morality is inadequate as a resource for solving environmental problems. The reason they give is that LG 4 of the light green position is false. In its place, they propose.

3. Living things or natural systems matter in themselves, that is, have independent moral statuses.

It is very interesting to witness the recent transition in the West from the anthropocentric, traditional, light green moral viewpoint to a new moral tradition of deep green, radical moral thinking. This new environmental ethic advocated our essential oneness with nature, where to be one with nature means to become part of "the great Self". This is evidently a cosmo-centric or eco-centric vision of our place in reality as this spirit is reflected in Indian philosophical tradition ever since the time of Vedas.

Let me now concentrate on how the deeper views of environmental ethics could be justified. There are lighter and darker green shades of environmental ethics possible as one may talk about animal-centred ethics, life-centred ethics and also we can talk about the rights for rocks and forests. Like the theory of human-centred ethics outlined earlier, the cosmo-centric view of environmental ethics is also made up of three components: a belief system, an ultimate moral attitude and a set of moral rules and standards. These elements stand in relation to each other in the same way that the three components of human ethics are interrelated. The belief system supports and makes intelligibly the adopting of the attitude, and the rules and standards to that moral attitude in practical life.

The belief system constitutes a philosophical worldview concerning the order of nature and the place of humans in it. This is called the cosmo-centric or deep-centred outlook on nature. When one conceives of oneself, one's relation to other living things, and the whole set of natural ecosystems on our planet in terms of this outlook, he/she identifies himself/herself as a member of the Earth's community of life. This does not entail a denial of one's personhood. Rather, it is a way of one's true self to include one's biological nature as well as one's personhood. From cosmo-centric perspective, one sees one's membership in the earth's community of life as providing a common bond with all the different species of animals and plants that have evolved over the ages. One becomes aware that, like all other living things on our planet, one's very existence depends on fundamental soundness and integrity of the biological system of nature. When one looks at this domain of life in its totality, he/she sees it to be a complex and unified web of interdependent parts.

This deep green outlook on nature also includes a certain way of perceiving and understanding each individual organism. Each is seen to be a teleological centre of life, pursuing its own good in its own unique way. A living thing is conceived as a unified system of organized activity, the constant tendency of which is to preserve its existence by protecting and promoting its well-being. Each living thing, human and non-human alike, is viewed as an entity pursuing its own good in its own way according to its species-specific nature. (For humans, this will include not only an autonomous, self-directed pursuing of one's good, but also the self-created conception of what one's true good is). No living thing will be considered inherently superior or inferior to any other. All are then judged to be equally deserving of moral concern and consideration.

The third component of such a theory of environmental ethics is the set of rules and standards to be morally binding upon everyone who has the capacities of a moral agent. These norms are principles that guide a moral agent's conduct with regard to how such an agent should or should not treat natural ecosystems and their wild communities of life. They are valid norms because they embody the attitude of respect for nature.

This cosmo-centric vision of environment or the idea of the deep ecology as propounded by Arne Naess has been the very spirit of the Indian tradition of thinking. In the context of Buddhist practical ethics, let me briefly take up the questions and issues concerning the respect for life in relation to animals for example. During the time of Buddha as well as during later debates, questions relating to such issues were discussed. Even kings were expected to provide protected territory not only for human beings but beasts of the forests and birds of the air. Buddha condemned deliberate infliction of torture to animals and killings.

There are four topics in the discourses, which are relevant to issues pertaining to the values of life: animal sacrifice, warfare, agriculture and meat eating. Buddha did not hesitate to condemn both the performance of animal sacrifices and pleasure of hunting. He also pointed out the futility of warfare. He prohibited the monks from joining the army and also from digging the ground, as in this process there was the danger of injuring insect life. But, regarding meat eating, he left it an open possibility that if one practices compassion, he/she would be inclined to practice vegetarianism. One can very clearly see the pragmatism and realism in Buddha's outlook, which provides useful resources for dealing with conflicts between human needs and moral ideals.

In the *Rig Veda*, the conception of *rta* is of great significance. It is the anticipation of the law of the *karma*—the law that pervades the whole world, which all goods and men must obey. *Rta* furnishes us with a standard of morality. It is the universal essence of things. It is the *satya* or the truth of things. The good are those who follow the path of *rta*, the true and the ordered. Ordered conduct is called a true *vrata*. The Vedas assume a very close and intimate relationship between men and God *vis-à-vis* nature. The life of man has to be led under the very eye of God.

The Indian philosophical thinking, is, in essence, spiritually-oriented, i.e., oriented towards the attainment of the highest good in life. Here, the highest and the deepest are and have to be the same. The highest in man's being is also the deepest. It is the highest because it is the most desirable; and it is the deepest because it is the ground of his being. This good is variously conceived in Indian thinking: as pleasure by the Carvakas; as an eternal becoming through ethical law by the Mimamsa; as conscious existence beyond the grip as *karma* by Jainas; as the *sunya*, selflessness and so beyond the reach of suffering by Buddhists; as unconscious existence by the Nyaya-Vaisesika; as conscious existence beyond the clutches of nature by the Samkhya-Yoga; and finally as the unity of consciousness, bliss and existence by the different schools of Vedanta. The cosmo-centric ethical perspective and our close affinity with nature is reflected even in *Ramayana*, *Mahabharata* and in most of the literary works of the Indian writers and poets.

References

Attfield, R. (1981), 'The Good of Trees', *Journal of Value Inquiry*, 15.

— and Belsey, A. (eds.) (1994), *Philosophy and the Natural Environment*, Cambridge University Press.

Benson, John (2000), *Environmental Ethics*, Routledge.

Benson, Andrew (1995), *The Ethics of the Environment*, Dartmouth Publishing Company.

Naess, A. (1973), 'The Shallow and the Deep, Long Range Ecology Movement', *Inquiry*, 16:95-100.

Passmore, J. (1980), *Man's Responsibility for Nature*, Duckworth.

Taylor, P.W. (1986), *Respect for Nature: A Theory of Environmental Ethics*, Princeton University Press.

Zimmerman, M.E. (ed.) (1993) *Environmental Philosophy*, Prentice Hall.

II

Religion, Culture and Environment

10

God, Human and Nature: Some Reflections from the Viewpoint of Christian Theism

Richard Howell

What, or who, is our ultimate concern. On what or on whom do we place the highest value of all? What is the dominant love, the supreme love of our life? Here, in this chapter, we will reflect on these questions from the Christian perspective.

Humanity has abused nature. This is witnessed by the phenomena of global warming and destruction of the earth's ozone shield that protects the earth, which has reached alarming proportion, with 30 per cent depletion by 1984. The acid rain, the defiled groundwater, lakes, rivers and oceans and water-borne diseases kill 25 million people in developing nations each year. Due to commercial use, the land for creatures and crops is reduced. This destroys land by erosion, stalinization and desertification. Deforestation each year removes 1,00,000 square kilometres of primary forest. About three species of plants and animals are eliminated from earth each day. The waste generation and global toxification results in distribution of troublesome materials worldwide by atmospheric and oceanic circulations.[1] The human and cultural degradation threatens and eliminates long-standing knowledge of nature communities regarding living sustainably and cooperatively with creation.

Years ago the scientist theologian Teilhard de Chardin said that he did not believe in the supreme effectiveness of the instinct of preservation and fear. It is not the fear of perishing but to live fully and beautifully that has fuelled sustained human endeavours and explorations.[2] What is needed is a passionate love for all forms of life, for matter and its mystery, for the future of the earth, and for the mystery and wonders of their Creator.

Nature has an economic value and the environmental crisis needs to be treated as an economic problem. Nature also has health value and the protection of the environment needs to be placed within the framework of preventive health care. Nature has recreative and aesthetic value too. We also need to be concerned about sustainable development for our next generations. Christian anthropocentrism does not place the autonomous person, nor the individual, nor a community, nor this generation but the whole of humanity of today and tomorrow at the centre. The reality of the ecological crisis has made people realize that a radical change in attitude is called for. The moral conflict caused by the ecological crisis necessitates a reconsidering of world-life and its values. The consideration of meaning and value arises out of the triad of meaning ascribed to God, human person and nature. Each of these three can serve as the primary reference point for ethical reflection.

The Human Person and Nature

The Bible depicts the person as a complex material-immaterial unity. Genesis describes the two-fold action of God that corresponds to the human person's physical and spiritual aspect (2.7). Materially, God formed Adam's body "from the dust of the ground". Concerning the immaterial dimension God "breathed into his nostril the breath of life". As a result of this two-fold creative activity, man became a living being. Soul is not a constituent part of the person, but the person or self in the totality of his being. The creation account also addresses the nature of the human person as created in the image of God. The persons do not have image; they are the unique representation of God. The human person is related to nature because he has been created out of the dust of the ground. But he is also related to God because he was created in the image of God. While dust relates person to nature and separates him from God, the image of God relates person to God and separates him from nature. Humans have certain unique capacities consisting of rational spirit or self-consciousness and self-determination by which humans stand in yet above physical nature. Therefore, a human person is considered the crown of creation. He is placed within a framework of theo-centric ethic. He has to give account of all his deeds before the Creator who provides. Many of the ecological problems take birth from social problems and economic injustices. The Christian anthropocentrism lives by obedience to God. Nature should not be governed by greedy and prideful motives;

a human person must rather live in all simplicity before God who foresees his need.

The Personal Absolute Incarnates on Earth

Since Copernicus' discoveries, the earth can no longer be considered the physical centre of the universe. The sun, as the centre of our solar system, seems unanchored in the expanse of endless galaxies. But, the cosmos is not a pointless field of diminishing energy destined eventually to burn itself out in the vastness of empty space. And, the value of human beings created in the image and likeness of God does not depend on the comparative size or location of their planet. Because God values human lives and the created world being, the transcendent sustainer of galaxies has come to earth in the person of Jesus Christ. The coming to earth of the infinite sustainer of all matter energy provides an absolute in nature. So Christians prize not only the personal absolute transcendent to nature, but also one personal absolute incarnate on earth. "For God was pleased to have all his fullness dwell in him" (Bible, Colossians 1:19). Jesus Christ, the eternal word became human flesh on planet earth to exhibit God's nature and teach God's truth. The desires of humanity for an ethical hero or a liberator have come to realization in Christ.

The word came into the world as Jesus in the flesh, and so the babe of Bethlehem is eternal. Jesus had two distinct natures—divine and human. The two sets of attributes are neither mixed nor confounded. No attribute of the one nature is transferred to the other. Neither is third hybrid produced. In Jesus Christ the transcendent God human and the physical nature came together: theo-andric-cosmos.

In contrast to totalistic relativist for whom history has no centre, no purpose and no abiding meaning, all history centres in the birth of God's eternal word. The creative word, without whom nothing exists, works in every age, but achieves his redemptive purposes in the fullness of time in history. Incarnation of Jesus Christ is the centre point from which we reflect on the role of the Christian in the world.

The Good Creation

At creation, God pronounced all things 'good' including the bodies of both male and the female (Gen. 1:31). Neither the matter-energy nor the desires of the bodies are inherently evil. Our bodies are not the fundamental source of our problems. With all the frustrations of our finitude, our deepest need is not to be free from our physical, spatial, temporal and energy limitations. We need not engage in bodily asceticism or self-indulgence. Our body abilities are like instruments that God's fallen image-bearers can use either for pride, greed, and just or to please our creator, who has our best interests at heart. Our fallen nature initiates unjust and immoral uses of our bodies. Although everything God created is good in itself, fallen people can use their bodies for evil as well as good. The sexual union of husband and wife is neither morally evil nor shameful but a beautiful portrayal of divine love when expressing partners' mutual faith, love and covenanted commitments to each other. People can marry to the glory of the Creator.

Some people imagine that the more spiritual ought to be vegetarians. Although created by God, animals are not God and are not made in the image of God. So, it is not morally wrong to kill them for food. Although creationists are free to eat meat. Jesus taught that food was not the most important thing in life (Matt. 6:25). Like any other material thing, meat may be taken if received with thanks-giving, eaten moderately according to God's word, and dedicated to purposes honouring God. But, if meat-eating would cause a person to stumble, creationists willingly give up the right to eat meat (Rom. 14:13-21; 1 Cor. 8:8-13).

Creationists have liberty to drink wine so long as it is used for good ends (1 Tim. 4:4-5). Although alcoholic drinks in moderation can serve good purposes, 'alcoholic abuse' expresses voluntary disregard to God's percepts. Doubtless, the Bible forbids drunkenness also because alcohol harms the abusers' minds and bodies, damages their home lives and interferes with their jobs. Our good Lord disallows drunkenness. The use of unprescribed substance shatters one's self-control and produces addiction, altered states of consciousness, drowsiness and develops into involuntary addiction. Alcoholism and substance abuse must be treated holistically both as physical and psychological disease and as a spiritual problem.

The Creator produced the limited resources of our planet for the benefit of all human beings, including coming generations. So, those who love God and their neighbour should live by a responsible conservationist ethic as Francis Schaeffer said in *Pollution and the Death of Man*. "When people take advantage of their special relationship by taking over God's creation as their own, using it for their own ends rather than God's glory, they have broken the covenant and are rebelling against God."[3]

The stewardship doctrine provides an answer to the ultimate question, "why should I take responsibility of caring for or preserving the natural order?" The answer is "because it is God's order". It is not for mere pragmatic reasons. If I love God, I will love what he has made.

Everything in nature has importance and worth because God made it for a just and loving purpose. And to its Creator we shall give account of our stewardship. So, Christian theists will not treat nature as if it were nothing or illusory, neither will they waste or pollute the resources for their own selfish gain. The responsibility of ruling over the animals and earth (Gen. 1:26, 28) is not a licence to waste them. Creationists will not avoidably waste resources, nor will they pollute the environment. They will use created resources altruistically as good stewards. Energy resources had a beginning and will have an end. Finite energy is subject to entropy. We ought not to rationalize unlimited wealth for selfish reasons. Stewards dare not rationalize irresponsible individualism, unaccountable enterprise, or unlimited exploitation of limited resources. This 'unaccountable theology' must be replaced by a concept of human dominion of the earth as stewards accountable to God. Energy users who are aware of the fact that they are accountable to the Lord of all and considerate of there own descendants, as well as the needs around the world, will not waste or pollute the planet's limited life-support system.

The Ultimate Origin of Creation

Our understanding of the origin of creation greatly influences our understanding of God, for an inescapable correlation exists between how one pictures the universe and how one views God. In the same manner, our concept of creation tremendously influences our understanding of the human person, the meaning of life and significantly shapes our attitude towards the environment and the larger issue of the utilization of earth's natural resources.

The question of ultimate origin reveals the difference between non-theistic scientists and theists. Scientists cannot uphold the thought of natural phenomena that cannot be explained. A pure empiricist does not even address the question beyond observation and repetition. No scientific observer was present at the beginning. Scientists play unscientific games when they speculate about primeval gases or energy. But, these had a beginning in the finite past and require more ultimate explanations. Three lines of evidence violate the scientists' religious faith and assumptions in natural phenomena, which support a beginning: (1) the outward motion of galaxies, (2) the law of thermo-dynamics, and (3) the life story of stars. Unable to discover the cause of the origin of the universe, the scientist should reconsider the exclusiveness of his method and naturalistic presuppositions.

Due to lack of evidence for an all-encompassing physical evolution, the hope is now set for an evolution of human consciousness. Fritjof

Capra, the advocate of new physics, initiates a whole series of social movements that all seem to go in the same direction. The rising concern with ecology, the strong interest in mysticism, the growing feminist awareness and the rediscovery of holistic approaches to health and healing are all manifestation.[4]

With the concern about the future of humanist has also come about a preoccupation with energy. Those in a desperate quest for lasting meaning think that their ultimate concern has to do with the finite energy of the universe. New consciousness and New Age thinkers with Paul Davies in *God and the New Physics* claim that the results of the new physicist are "more like mysticism than materialism".[5]

The modern physical discoveries have been valuable to a certain degree, yet the energy of the universe is not the unlimited personal God of the Bible: (1) The energy of the universe loses its usefulness, God does not. (2) The energy of the universe had a beginning and will have an end. God does not end. (3) The energy of the universe is impersonal, God is personal. (4) The energy of the universe is amoral, God is moral. (5) The energy of the universe can be manipulated by human engineering, God will not be manipulated by physicists or mediums.

The human persons are not mere field of energy. The hypothesis that reduces humanness to impersonal energy dehumanizes persons. If there are no personal agents, who communicates with whom and who is responsible for moral conduct? And to what? To what field of energy shall we assign responsibility for the view that person is nothing but their actions or relationships? The loss of life's meaning will quickly follow the loss of cosmic meaning. Only in pantheism are all persons ontologically one. In theism persons are distinct beings created as unique individuals for fellowship with a personal God. The loyalty of persons to one another in families, societies and nations can reflect a significant solidarity of values, commitments and relationships, but not a sameness of being. The Christian scripture recognizes not only humanity's inner spiritual reality but also outward reality as bodies. Materialism reduces all human spirituality and physiology to physics and its epiphenomenal by products. Idealism tends to regard only the inner soul or spirit real and the body unreal or illusory. Given the reality of human spirits and their consequent inherent worth, personal relationships are of great value. The worth of personal relationships is enhanced when we recognize the inherent values and rights of persons. Knowing what was in humans, Jesus Christ said that a person's soul is of greater value than the whole world. Why? Because persons have inalienable value. Created persons always depend on God for life and breath and must always be seen in that dependent relationship.

In the sphere of 'deep ecology', nature evolutionistically and ecologically defined functions as a central reference point. The proponents reject the view that conceives human person as the centre of the cosmos and advocate that human person needs to think and live from the standpoint of the biosphere, the third entity in the triad of meaning. The earth is viewed as a self-regulation organism. This leads indeed to depersonalization of God and personalizing of earth. Since God as the law-giver is rejected, nature becomes the basis for ethics. In Christian thought, theology is the foundation for Christian ethics and hence indivisibly connected. The foundational question is: which God or god is at the centre and how should this divine being be described in anthropomorphic or biosphere terms Christian theology chooses biblical revelation to describe God's nature.

In the beginning God created (out of nothing) the heavens and the earth (Gen. 1:1). The opening verse of the Bible immediately confronts the reader with the eternal and transcendent God who is there. God is the sovereign Lord. Scriptures describe him as the other, and as Creator; He is radically distinguished from his creation. The Bible gives a close connection between God as King and God as Creator. He is the mighty sovereign who knows how to control the chaotic powers of nature.

Also, the relationship between Christ and the cosmos is one of the dominant King-Creator over against his creation. Ecological ethics must start from God as the transcending authority. Alienation from God as 'law-giver' is also alienation from nature, which he created. If people set himself against this basic truth, which happens in earth-centric movement, another will replace the one authority. The value of nature must build up from the relation to the holy Creator. God is for the Christian ethics, the only one which can give objective worth.

God is self-existent and eternal; the universe is dependent, having a beginning in the finite past. And to say nature is self-directed attributes intelligence to the non-intelligent. Apart from a transcendent God, the impersonal physical world has no mind or will and hence no wisely chosen 'direction'. Any testing meaning of historical existence has lost its foundation if the case for an objectively real God and accountability are not accepted.

The Fall of Creation

Since the fall of creation, nature has been characterized by a value-based paradox of worth and non-worth. The present non-worth of nature and the rift between Creator and creation is brought about primarily through sin and death. In eco-theology, death is not considered an abnormality but an ecological necessity. Sin in eco-theology has no effect on ecosystems but remains limited only to social systems, in particular to the dominant structures in society. The presence of death, sickness and chaos brings about a deep rift between Creator and creation. Nature is subjected to transitoriness and death is not biological evidence rather the last enemy that must be conquered. Given the reality of chaotic and even demonic powers in nature, the idea of 'ecological balance' is an illusion.

The Ecological Expectation

In spite of the fall, nature retains value: (1) The nature still declares the Glory of God. (2) The incarnation and the bodily resurrection of Jesus Christ proves that God has not abandoned nature. Through incarnation of Jesus Christ, creation's value is reaffirmed, and by the resurrection of Jesus Christ, the basis has been provided for a material eschatology. (3) The nature belongs to God, the Creator-ing, and Holy Father and therefore has value because God owns it. (4) Nature is God's garden, and through this God provides for the needs of humanity and through which he blesses and judges. (5) The relation between God and nature is also seen within the context of plan of salvation. The scriptures speak of the eventual sanctification of nature. Paul in Romans indicates that nature itself will participate and be liberated from its bondage to decay (Romans 8:21), i.e., from futility, vanity, entropy, decay and death. God of his own righteous will placed all creation, including humanity, under judgement. God has judged it by his righteousness and will redeem it by his grace. The redemption of nature is a sub point of the redemption of the Christian and should be viewed from an anthropocentric perspective.

The Enlightenment: Human Person

The modern human subject is no longer an enlightened human being. So, the 'postmodern' thought is involved in overcoming the autonomous human subject of modernity. The task appears to be a decent ring of the human. Postmodernity, in spite of its rejection of autonomous human subject of modernity, continues to share modernity's anthropocentric horizon. For postmodernity and modern thought, the human remains the contingent, nevertheless the final horizon of thought and action.

Christianity, like postmodernity, sees the need for a decent ring of the human in order to overcome the autonomous subject of modernity. But Christianity unlike postmodernity seeks to achieve this new wholesomeness of humanity by way of the light of the transcendent logos, Jesus Christ. What Christianity calls for is an openness to the Trinitarian God as revealed in Jesus Christ as the revealed conceptuality for understanding reality and experience and to display the concrete meaning of a human person's fundamental relation to God and nature and show how this relation serves as the basic horizon for thought and action in the service of true faith.

The Nature of Work

How should Christians work? "Whatever you do, work at it with all your heart, as working for the Lord and not for men" (Col. 3:23; Eph. 6:7-8). The ultimate owner of the universe is distinct from it, and so private ownership of land and resources is never ultimate, but a temporary stewardship under the Creator. God lends us some property and its resources temporarily for use in efficient service of others. Christians enjoy each day of life on earth, whether they are working or retired. The Creator ordained that we work. "Six days you shall labour and do all your work" (Ex. 20:8). The ultimate stimulus for the work ethic comes from God's example in creation. "For in six days the LORD made the heavens and the earth, the sea, and all that is in them (Gen. 1:11). God created humans to share not only his fellowship but also his work. More important than what we do for a livelihood is who we live for. Whatever Christians do they work at it ultimately for the glory of God (1 Cor. 10:31).

Lewis and Demarest write: "What is work? Negatively, work is neither an evil nor a punishment for the fall. Affirmatively, work is the use of God-given energy as a means to achieve a good end. We misunderstand work if we think it is an end in itself, like play. After the fall work may be more irksome, but it still benefits others and provides purpose and fulfilment. In losing ourselves in service for others we find ourselves. By way of definition, work is the expenditure of mental and physical energy to the best of our trained abilities in making quality products or doing needed services that we "may have something to share with those in need" (Eph. 4:28)."[6]

The central goal of work is to provide the best possible service of product. A Christian's daily work is primarily rendered to the Lord and not humans (Eph. 6:7) and it is to be done with integrity (Col. 3:22-24).

The lowliest task on earth can be viewed as a service for others when we love God who gave us the gifts for our work. When we love others for whom and with whom we work, no amount of success, fame and private accumulation of goods will keep us from making a commitment for others' safety, health, clothing, housing, protection, justice and education. And, in so doing, we find ourselves fulfilled. With servant's heart like our Lord's we live not to be served in high positions but to serve in humility (Mark 10:43-45).

Conclusion

Since God is the creator of matter, it has been given dignity and status before God. It is seen declaring the glory of God. However, it is in the person of Jesus, his incarnation and historical service, and in his death and resurrection that the ultimate greatness and wonder of matter stands revealed and affirmed. So, we have a Christ-centred universe shaped by God, through Christ and moving towards him as towards its pole of completion. He, therefore, is the Alpha and Omega of the cosmos. Hence, it is God's plan to reconcile everything to himself in and through Christ. The mystery of God's purpose which he will act upon when everything together under Christ as head, everything in the heavens and everything on earth (Bible, Ephesians chapter 1).

Those who love God increasingly detest evil. In the face of social injustice to the poor and powerless, the godly hate oppression, discrimination, selfishness, greed, covetousness, and all the contributing evils. We continue to look forward to the moment that God himself makes all things new, meanwhile we must take our stewardship seriously for our fellow person and nature both now and in the future.

Notes

1. Evangelical Christianity and Environment Joint Declaration by Ethics and Society Study Unit of WER-TC and Au Sable Institute, 1993 WEF Theological Commission P.O. Box 94 Choong Jong No, Seoul, Dorea, pp. 120-650.

2. Teilhard de Chardin, *Building the Earth*, Wilkes-Batte, P.A: (Dimension Books, 1965), p. 36.

3. Francis Shaeffer, *Pollution and the Death of Man* (Wheaton: Tyndle, 1990), pp. 49-50.

4. Fritj Capra, *The Tao of Physics* (Boulder, Colo: Shamballa, 1983), p. 8.

5. Paul Davies, *God and the New Physics* (New York: Simon and Schuster: 1983), p.vii.

6. Gordon Lewis and Bruce Demarest, *Integrative Theology* (Academic Books Grand Rapids Michigan), Vol. 1, p. 63.

11

Glimpses of Environmental Ethics from Indian Scriptures

P.K. Bajpai and Nitish Dubey

There is an impressive account of religious scriptures that emphasizes upon the environmental ethics and honour of various forces and components of environmental aspects around us. It is difficult to present the entire text concerned with such aspects and in a limited space it is difficult to mention all related references.

However, we can seek some important views expressed in *Srimadbhagwata Gita* and a few from *Shukla Yajurveda*. The entire Vedic literature, throughout its text, is marked with worship of nature. Most profused prayed deities in the Vedic text are *Vishnu, Indra, Sun, Fire*, etc. But, the subject of the sixteenth chapter of *Yajurveda* is largely *Rudra*, the multifarious but principally the destroyer authority in the trinity of *Adi Devas*.

For the present consideration, it is proposed to begin with ideas given in *Gita* and then to proceed back to *Yajurveda*. The third chapter of *Gita*, named as *Karmyoga*, deals with human duties and ethics that would render a good sustainability of human existence and environmental consistency going together. *Yajna* is the subject of verses 8 to 25 and Lord Krishna dictates the various do's and don'ts in that context. We will take the relevant verses in the same sequence as given in original text but briefly, for limitations imposed in relation to space and time.

Verse 9 states that the people who are engaged in acts other than those concerned with *yajna* fall into bondage of *karma*, therefore, one should act for *yajna*, without an attachment. It also tells us that only duties or those act, which are desirable, are to be done and not otherwise.

Verse 13 advises that the sustenance obtained as left-out from *yajna* is consumable, because those who cook to feed their physical selves are engaged with sin only.

Verses 14-15 are most significant in this context. Lord Krishna tells to Arjuna an account of ecological cycle in brief. He speaks:

"People are product of food, food is product of rains and rains are product of *yajna*, while *yajna* is the consequence, and originates from the *karma*. The *karmas* have their origin in Vedas and Vedas came from Eternal Being, the God, and it indicates that this eternal being has its establishment in *yajna*."

The above account of 14-15 verses, if read in reverse order, would mean that God gave the code of conduct to man through Vedic text and acts rendered accordingly amount to be a *yajna* and that *yajna* gives rise to clouds, and the water showered by them generates food that gives sustenance to human populace. Now it is obvious that any act, which is contrary to the code, shall not result into the merits supposed to be obtained from *yajna*. In other words, whatever we do is a contribution to *yajna* or to say that each act whether deliberate or indeliberate or involuntary or voluntary are the offerings to the *yajna* of life. If offerings are proper, they must result into correct or classical *yajna* but an improper act will pollute the *yajna* and then it will result in rise of clouds and ultimately will bring food to scarcity force starvation on larger section of humanity.

The question arises what to do that may be proper, if we do not consider the 'don'ts' as well? The answer lies in the sixteenth chapter of *Yajurveda*, also known as *Neelsukta* included as fifth chapter of *Rudrashthadyayi* (*Shukla Yajurvediya*). The said chapter devoted to varied forms of Rudra, as postulated by the seers of that Veda. Each *richa* is an expression of honour to one or the other form of Rudra. We will now go into a brief account of various forms of Rudra which deserve honour.[1 & 2]

"The lord of all directions, the trees adorned by green leaves as hair of man, the caretaker of animals (cattle), the Rudra having complexion of young yellow leaves (17), Lord of cereals, Lord of all life forms, protector of cultivated fields, Lord of trees, Lord of medicinal herbs (19), the products of fertile land (33), born in forests, and among climbing herbs (34), the dweller of large canals, lower reaches of hills and mountains, lakes, rivers, ponds, wells, ditches, clouds, thunder, rains, dry and draughts, in tornadoes, and Catastrophes (37-39), the lord of beasts, and trees with green leaves (40), those produced in sand, springs, small stones and pebbles, stagnant waters, barren deserts, in the herd of cows, difficult lands and caves, in the dry and green wood, in dust and in fine dust land and sea fire (43-45)"—all are offered namaskar so that the men and cattle are blessed to keep good health and without restlessness. Further, the verse 58 reads that among trees those with yellow leaves are the Rudra who have blue throat. The story of blue throat (Nilgriva) is that Siva by way of drinking poison and retaining it in his throat got that blued. Now by way of reading between the lines it can easily be deduced that probably the seers knew. The pollution absorbing capacity plants, specially the young plants that have to undertake photosynthesis at a relatively larger scale, so that to save food to built their bulk after spending energy for maintaining the life."

The above quotations tell us that almost no form of life whether microscopic bacteria or protozoan living in soils and mud or larger mammals are there, which has been spared from being regarded as expression of Rudra who is also the lord of forests and beasts. The *mantras* are meant for requesting the blessings and kindness of Rudra and hence anything that is under his lordship cannot be subjected to dishonour.

As said earlier that according to Krishna the dictums of Veda decide and show the correct path of life, these forms of life and ecosystems sustaining them are honourable and once they are honourable they cannot be misused or overused and if so the environment would remain undisturbed to a larger extent and such a balanced environment should certainly bring good rains down to earth for good harvests to be reaped.

So to say, India does not require an import of ideas and codes of environmental ethics, as it has a rich heritage of that kind in its own traditions and scriptures. What is required is simply to interpret and understand the ideology, already given, in present context of environmental protection.

It would not be out of context to take the first verse of 40th chapter of *Yajurveda*, i.e., of *Ishavasyopnishada*. This *richa* or verse is of utmost significance as regards our present context. It tells us:

"The entire creation (all existence) is brought with God, regardless of visible or invisible, wherever it is a trace of creation. Whatever is given by that Being is to be consumed without any attachment. Do not be greedy as after all whose wealth is this?"

The verse includes some very important statements that need a little explanation:

Taking it as a condition—the world (or entire creation) is full of God, it may further be said that entire creation is within God and vice versa. When God and creation, both are so intimate there may be an apparent distinction between two or an illusion of duality may exist but reality is the union, i.e., one without the second. Following this implication, all living beings are but a manifestation of God and hence within their individual capacities are obliged to maintain a balanced living.

Whatever is parted away is to be consumed as sustenance, implies that exploitation of natural resources by man is an act that must be kept out of question since man himself cannot exist without God Himself and self cannot and should not exploit self.

Whose wealth is this? The raised question has our answer within—no one's wealth is this. The natural resources or environmental resources accordingly may not be treated as propriety of an individual. It is a shared and to be reasonably shared wealth. Hence, greed is impermissible. The seer knew that a permission of greed and possession would amount to unwarranted interference and disturbances that would ultimately lead to shattering and collapse of the system.

Much of philosophical commentaries are available from different authorities, which had attempted the Prasthanatrayi, but possibly none has taken the environmental concern into consideration.

The present-day attitude of mankind, as West has led us in scientific endeavours and achievement, largely accords with occidental ideology that all creation is for being used by man and when such a concept becomes leading one, the question of rights to use becomes dominant over fading shadows of duties towards nature or in Vedic context the Creator and creation relationship.

The last verse of *Ishavashyopanishada* speaks of such weakness of human nature, thousands of years before the occidental ideology came to mind. *Hirnimayen Patren Satysapihitam Mukhain*—the opening of truth is covered with a golden lid. The gold has always been the source of inducing greed and greed is very likely to weave away the mind from probing into truth, i.e., what is behind the golden lid. For a good long time of over past two centuries man has seen only the golden lid and has not looked into the truth of survival. It is only when most of the lid has been concerned, destroyed and the truth of survival has been exposed to uncountable types of dangers, the significance of protection is hammering us.

The most ironical situation is that creation is facing a threat from within because we did not realize our life as a pious *yajna*, we did not follow the concept of Creator-creation relationship and beyond all, man stood away from nature and took the latter as a means of overutilization, forgetting the fact that man is within nature and nature is within man, the two are, like all other existences, the inseparables. To attempt to harm nature is an attempt to harm ourselves, may be indirectly or in disguise.

Notes

1. The quotes are translations from *Shukla Yajurveda* edited by Dr. Ram Krishna Shastri on the basis of commentaries of Uvatacharya and Mahidhan, Chaukhambha Vidya Bhavan, Varanasi.
2. The numbers in parentheses indicate the number of verses, i.e., *richa*.

12

Environment and Vedic Literature

Ranjay Pratap Singh

Environmentalism as a basic ideology is an explicit element of the history of India from time immemorial. Living in mutual harmony with nature has been a part of India's cultural as well as survival heritage. To survive in and with an environment of such ecological and cultural diversity, traditional Indian societies evolved a prudent system of resource use that was entrenched in a deep-rooted ethic of conservation and sustainable resources management. Ancient Indian texts are a rich source of information and insight on the historical root of Indian environmentalism. The *Rigveda* is the most important of these; it contains invocation addressed to various gods and goddesses relating to different personified powers of nature. The *Atharvaveda* is especially important for the history of science in India. It describes the magical properties of several plants. The *Yajurveda* lays down ritual and methods pertaining to performance of *yajnas* (sacrifices).

The earth, according to the Vedas, is not only for human beings to enjoy but also for all other bipeds and insects and other creatures (*Atharvaveda*, 12.1.15). The most important aspect of Hindu theology is the belief that the Supreme Being actually gets incarnated in the form of various species (*Srimad Bhagawatam*, Book 1, Discourse III: 5). Indeed, ancient Hindu societies understood God and nature (*prakriti*) to be one and the same (*Mahabharata*, moksa, 182). It was the original human instinct to survive that of 'nature as divine' began in the hunter-gatherer stage and continued in the pastoral and agricultural societies. While these societies indulged in the clearing of forests and killing of wild animals, they did manage to retain an awareness of the need for ecological prudence and preserved many practices of the earlier tribal period. The importance of cattle in Vedic period was not merely economic but also ritualistic and symbolic.

Atmosphere surrounding us is called *paryavarana*, which we consume directly or indirectly. Generally, green natural surrounding around us is called environment (*hari bhari prakrati ka avaran* means *parayarana*). Here, *Hari* means '*sujala, sufala, malayaj sheetala shasya shyamla*', i.e., mother earth and *bhari* means rich fauna and flora. Atmosphere gives name and glory to earth as a living world. In the entire solar system, only our earth provides favourable atmosphere for the growth and development of plants and animals. Besides human beings, animals, bird, trees and shrubs are all gifts of nature or atmosphere. Since long, these creatures are living on earth in a very natural and disciplined way. This balance of nature is not only necessary for our culture and civilization but for our proper physical and mental development. If we are not able to breathe even fresh air and drink pure water then all other benefits of modern development becomes meaningless.

Importance of religious beliefs regarding environment conservation cannot be questioned. If we look into the basic concepts of predominant religions, we find that Islam preaches brotherhood, which is a universal human value. In the same way, Buddhism preaches eight-fold path to avoid sorrow and Jainism preaches non-violence and avoidance of unnecessary collection of property (*aparigrah*). By worshipping the natural forces Vedic religion opens the path of fulfilment by getting the knowledge of almighty God and thus attaining salvation. Other *Vaishnava* religions also worship God with the salutation of nature. Although Jaina religion has shown futuristic attitude towards environment conservation by imposing element of life on the physical objects, Vedic philosophy goes one step further by imposing divinity to physical objects. It has turned them into objects of worship. Vedic rituals like *yajna*, etc., have been designed to fulfil the objects of environment conservation and maintain the balance between Pinda and Brahmand (that is, man and solar system).

Feeling of unity amongst diversity is basic principle of Vedas. This feeling not only extends to all human beings but also to other living creatures and subconscious world. Vedas accept one God as the root of many gods *(ekam sad vipra bahudha vadanti)*. In *Rigveda*, origin of four *varnas* by God's mouth, arms, thighs and feet; and origin of moon, sun, air (*prana*); and fire from his mind, eye, nose and mouth is the environment-based theory of creation of universe. In the same way, it is found

that after *Purushmedha Yagya,* God created the universe. He created space
by his navel, higher worlds from upper part of body, earth from the feet
and directions (*disha*) from the nose. Thus, it is clear from the Vedic
theory of creation that one God creates gods, human beings, animals,
birds, trees and plants, but also other constituents of environment. This
divine vision towards environment has imposed divinity on all its
constituents (*Ishavasyamidam sarvam*) and formulated the concepts of
family relations and prescriptions.

During Vedic times, human lifestyle was totally dependent on
nature. People believed that nature possesses wonderful powers, which
can deliver good or bring harm to them. Vedic philosophers believe in
the theory of 'as the man, so the universe' (*yatha pinde tatha brahmande*).
They believed that human life is comparable to the whole universe and
man does not live in isolation but leads a collective life. Various prayers
have been composed in Vedas to please these small or big powers of
nature and seek their blessings for our pleasant living. Vedic people were
worshipping *parjanya, sama, savita, upas,* fire, air, water, animals, trees
and vegetation. Keeping this perspective in mind, we find that Vedic
worship is nothing but environmental worship. This worship of smallest
elements of nature is not due to fear from them but due to our gratitude
towards them.

Acceptance of sentimental relations between man and environment
is a hallmark of Vedic concept. To love and continuously try for proportional development of the soil, hills, rivers, animals, birds, trees and
plants of the environment where we are born and are nourished by its
food, water and air, are human expressions of Vedic concepts. Feeling of
love and attachment towards environment, expression of belief,
gratitude and sacrifice in special circumstances for environmental conservation are true human sentiments. If we feel oneness with all the
constituents of environment of a particular area, feel happy if it prospers
and feel sad if it degrades; this is the test of our environmental
consciousness. This is the Vedic concept towards environment and its
constituents.

Vedic concepts can be easily found at the root of human concepts
towards environment. Vedic civilization is very ancient. It was spreading
the knowledge when other civilizations were at the nascent state. In
Vedic literature we find highest form of human sentiments towards
environment and *Puranic* literature is also its embodiment.

Vedas and other Upanishadic literature gave broader meanings to
feeling of unity in diversity. This feeling was extended not only to divine
but also to all living and non-living creatures. Upanishads explained that

since God is present in all the things of universe, we must not have a consumeristic attitude towards them. As Lord Krishna said in Gita— 'who looks everything in God and God in everything', is important in this respect. Thus, ancient sages concentrated not only on human beings but also on all the living, nonliving, abstract and mysterious things of universe, many of them are not visible to us. Many concepts of Aryan lifestyle and philosophy despite being old are still relevant and useful today. Thousands of sayings like: *'Ekam sad vipra bahudha vadanti'*, *'Isha vasyamidam sarvam'*, *'Vasudhaiva kutumbakam'*, *'Sarve bhavantu sukhinah'* have reflected in the ideas like unity in diversity, live and let live, simple living and high thinking. Not having consumeristic attitude but feelings of non-violence, pity and compassion are in the background of these thoughts. Thus, human thought towards environment has been inspired by these Vedic principles, which are refined as well as broad in meaning. Generally, sky, earth, sun, moon, air, fire, stars, forests, hills, rivers, oceans, animals, birds, trees, plants, etc., are included in the environment. Ancient literature is full of divine and fellow feeling towards these elements. In the Indian folk culture, mother, mother earth and mother cow has been given place of prime importance. Every Indian feels lifelong obligation towards his country due to the feeling of *'Janani janma bhumisca swargadapigariyasi'* and remains prepared to sacrifice himself at the time of need. Thousands of persons laid down their lives chanting the slogan *'Vande Mataram'* during freedom struggle is a testimony to this fact. Mother earth nourishing thousands of creatures besides human beings is also depicted as the wife of God Vishnu. Thus, every Indian prays to her in the morning and begs pardon for touching her by one's feet:

Samudra vasane devi parvat stan mandale
Vishnupatni namastubhyam padasparsha ksamasvame.

We have relation of mother and son to earth that is opposed to the western concepts of consumer and consumption. Since it is believed that son may be bad but mother cannot be bad, no other relation can be so pure and pleasant than the relation between mother and son. Besides this, while the blood relations attach all the human beings, mother earth attaches all the creatures by her earthly smell alone. Vedic sages pray to this loving mother-goddess earth: that you are our mother, we are your sons and we salute you; Mother Earth, we harm you by many acts, these injuries may be cured quickly. In the same way Dyau (sky) is our father. Father Dyau and mother earth nourishes all the creatures. Vedic sages treat Marut (air) as medicine for the heart. They thought that pure air could prolong the lifespan of an individual.

Thus, Vedic sages treating air as father, brother and friend pray for long life from them. They thought the *Amrit* (life giving mysterious substance) is at the heart of pure air, so they pray for *Amratatva* (never ending life) from them.

Vedic sages were aware of the beneficial aspects of 'Savitra Dev' (sun), as a component of environment. In his prayers, this quality of removing diseases and misery is aptly mentioned. Sun gives life to creatures by his light, so he is also father of all the living beings. Sun removes impurities and purifies all the things.

Highest numbers of *sutras* (prayers) are composed for Indra and second place is given to *Agni* (fire). According to sages, taking *yajnik* offerings to different gods, creation of clouds, movement of air and origin of medicines all are due to fire. Whole Vedic literature is full of prayers to such beneficial fire god, praying for blessing from him.

Great poet Kalidasa has treated eight important constituents of environment that is earth, water, air, sky, fire, sun, moon and Ritvij as eight-fold statue of god Shiva and prays for blessings from them in *Abhigyanshakuntalam*. In *Malavikagnimitram* also, Kalidas requested these eight forms of Shiva for nourishing the universe.

On one side, Divinity has been imposed on *Dyau, Savita, Sama, Marut*, earth, etc., on the other they have been treated very normally with other celestial bodies. Many folk tales regarding mythological as well as ordinary characters of planets, the Sun, Moon, Mars, Venus, Saturn, Mercury, Jupiter, etc., are found. Many stories about *Rahu-Ketu* are also found. Human sentiments have replaced natural phenomenon like solar and lunar eclipse at the folk level in a very interesting way. Through these stories earth has been depicted as mother and wife of Vishnu, Dyau and Savita as father, *marudgana* (air) as friend and brother, mars as son of earth, moon as treasure of medicine and mother's brother (*chandamama*), Jupiter as teacher of gods, Venus as teacher of demons and father of Devayani and *Rahu-Ketu* as harmful planets.

Indian folk culture does not attach human sentiments only to earth, water, air, sky, fire, sun, etc., but also extends them to rivers, ponds, hills, animals, trees, and plants as well. This is a continuing tradition. Ancient Vedic sages have given a human shape to earth and all these are its integral parts. As we know, according to Vedas, first of all medicines appear on the surface of the earth, prior to this the earth was barren and coverless like the back of tortoise (*kurmprishanibha*). In those circumstances, origin of animals was not possible. Medicinal plants are like the hair of earth. As hair is beneficial to human beings, so are plants to the earth also. In the same way, rivers are life giving blood vessels of the earth. Indian people have attached human and divine feeling to rivers. All rivers flowing in India like Ganga, Yamuna, Saraswati, Narmada, Sindhu, Krishna, Kaveri, etc., are treated as goddesses. Every Indian feels his pious duty to take a dip in them. Due to this feeling of divinity, people while bathing at all these places, have more or less the same feelings.

Many folk and mythical tales are attached with all the rivers of India. Thousands of interesting compositions and myths are found regarding them in ancient literature especially in *Puranas* and other Upanishadic literature. This has not only strengthened feelings of worshiper and the worshiped, i.e., between human beings and rivers but has also cemented traditional family relations. Due to this prevalent sympathetic thought, polluting rivers, ponds and wells are reprehensible amongst Indian phyche. It is believed that *Agni* lives in water, which is also a symbol of *Varun*. Thus, not only naked bathing but all other acts denigrating divinity and disrespectful to rivers were prohibited. These religious introductions are still effective in controlling river pollution. Provisions were made to remove the sin of water pollution by conducting *chandrayan vrata* as told by *Nirnayasindhu* of *Bhavishyapurana*.

Thus due to the feeling of divinity, Indian code of conduct has provided stringent measures to prevent distortion of even the components of environment. It will be difficult to find such kind and humane feelings in any other culture. Smritis or code of conducts proposed by ancients—*Manu, Yagyavalkya, Harit, Gautam, Katyayana, Naral and Vashishta* have set milestones in this respect.

Hils, by giving birth to rivers, providing prosperity are also important constituents of environment. Due to this reason, hills are also treated as divine subjects for worship. These natural relations between rivers and hills get reflected in literature, which became means of spreading human sentiments. This resulted in the acceptance of most river-hill, river-forest, river-sea and river-river combination areas as Tirthas, meaning places of divine powers and places of worship (*Kumar Sambhav*, 58/11, 12).

Protecting these places of motherland is our solemn duty. Due to these sentiments towards such natural places, Indian soldiers have laid down their lives in defending inaccessible terrains of Kargil in July 1999.

Kargil victory has proved once again that we will not allow encroachment of even the smallest part of our motherland.

It is said that animal and forest is life, which means that on earth neither forest can survive without animals nor animals can survive without forests. In the absence of one, existence of other cannot be imagined. Feelings of humanity and divinity towards trees and plants are found everywhere in the Vedic literature. Inspired and obliged by plant's quality to benefit others, ancient sages have praised them freely. *Vrihaspati* has been treated as inventor of medicines and *Som* as master of medicines. Due to imposition of divinity on medicines, sages have composed many verses of prayer for good health and hundred-year life from them. Sages pray that medicines like mother give good health to us. They wish them to prosper with many leaves, flowers and fruits to benefit all the sick persons (*Rigveda* 1,17,3-5).

Sage Yask treats medicines as source of energy and destroyer of pollution. According to him, medicines are able to end burning sensation and can remove pollution means. According to *Atharvaveda* the whole world is based on a circle named Chandas, which includes water, air and medicines. Thus, like air and water these divine medicines are also essential for the survival of mankind. Two energies, namely, *antardhi* (internal energy) and *paridhi* (external energy) are found in every particle of universe. Life remains secure in terms of the balance between these two and becomes endangered when this balance gets disturbed due to environmental pollution (*Atharvaveda* 8.2.25).

Medicines and vegetation are called Rudras in *Yajurveda*. These Rudras are of two types: Shaiva, i.e., the beneficial form and Rudra, i.e., the harmful form. All the constituents of nature remain Shiva in conditions of balance and become Rudra due to imbalance. Rudra is also known as poison taker and life giver, i.e., Shankar. These trees and vegetation also take poison in the form of carbon dioxide and bless with life-giving oxygen. Due to imbalance of carbon dioxide rudra becomes harmful. Thus, serving the vegetation is serving the Rudra god (*Yajurveda* 16.49).

This shows that our ancient sages and intellectuals have recognized threefold importance that is birth giving, nourishing and destroying from trees and vegetation thousands of years ago. By a well-decided plan they have made them points of worship much central to our religious, social, cultural activities and our daily lives. Imposition of divinity in trees and vegetation is a result of this thought process, which is relevant and useful in the present time also.

Ancient sages also tried to develop sensitivity towards birds and animals and desire for conserving them amongst people by making them equal to gods, symbol of gods or carrier (*vahan*) of gods. Due to this, these animals are found intermingled with people in many forms. Indian folk traditions gives religious importance to cow as mother, elephant as Ganesh, ox as Nandi and Shiva *vahan*, lion as Durga *vahan*, donkey as *Sheetala vahan* and rat as Ganesh *vahan*. In the same way amongst birds Garud (kite) is respected as Vishna *vahan*, Hans (swan) as Brahma *vahan*, mayur (peacock) as Kartikeya *vahan* and ullu (owl) as Lakshmi *vahan*. According to regional beliefs, snakes are worshipped at Nagpanchmi, horses and *Neelkanth* birds are worshipped at Vijayadashmi or Dashahra and cows at auspicious occasions. Due to human feelings people believe that watching *nevla* in morning and *Neelkanth* at Vijayadashmi are good. In the same way, at the time of departure, watching elephant house, cow, ox and love bird is believed to be good. At many places traditions are found regarding feeding of animals with flour to ants, flour balls to fishes, fried gram to monkeys, grain (*annagras*) to cows, *pakodi* and *vada* to kites, grain (*annabali*) to crows and food (*kaura*) to dogs.

Expression of family relations with trees and vegetation are also based on Vedic literature. Ancient sages have not only imposed concept of divinity on trees and vegetation but also created a space for the expression of human sentiments. In the Vedic literature, feeling of mother, father, son, friend, brother, etc., are prevalent among human beings and plants. Medicines, which nourish the world, protect us like mother (*Rigveda* 10,97,4).

Aranyani Sukta, showing human concerns towards trees and vegetation is important from the point of environment conservation. Here *Aranyani* has been called as mother of all creatures. Whole Vedic literature is full of these feelings. Embodiments of family relations towards trees and vegetation are found abundantly in religious and folk Sanskrit literature. Due to these relationships, believing tree as *Dharmputras* or son became especially popular. Most of Sanskrit literature is full of these sentiments towards conservation of environment. It is believed that like *yagyas*, trees also help in attaining salvation.

Thus, with the aim of conserving natural elements to keep the balance of universe in a proper shape, ancient sages in their extraordinary wisdom, first imposed concept of divinity to them, then established sensibility of family relations and finally attached concepts of *pap-punya* (sin-sacred) with them. It means that conservation of divine or mother, father, brother, son like constituents of environment were declared sacred act and their destruction a sin. Whole ancient literature is full of these concepts of (*pap-punya*) vice-virtue towards natural elements.

Thus, Vedic, religious and Sanskrit literature are not superfluous hypothetical expression of divinity and family relations with constituents of environment, but it is extreme expression of human relations and a practical scientific foresight of our ancient sages. In the modern times, mankind totally gets embedded in materialistic ambitions, selfish motives and become insensitive towards ecological surroundings. Western thought processes has not only evaporated boundaries and sanctity between relations but has also deformed them. Due to this approach animals, birds, trees, vegetation, etc., have become only consumables and means to attain selfish motives presented by nature. These feelings are leading into cruel destruction of natural environmental elements and unending race for physical possessions. Due to this vision of extreme materialism, man looses all emotional relations with animals and birds, which now became only objects to devour or decorate. This vision also leads to negation of family relations. Wife, friend servant, son, etc., are considered only to fulfil selfish ambition or to satisfy physical and animal instincts. Due to this narrow vision and insensitivity towards relation's families, societies and nations are facing problems and everywhere violence and anarchy are prevalent. In these diverse conditions besides applying codes of conduct strictly, popularization of ancient concepts towards environment is also necessary. This may help enlighten people's scientific, familial and ethical feelings towards environment and reduce insensitivity in them. This experiment for conservation and enrichment of natural forces will definitely succeed. Our ancient but futuristic sages expressed their feelings about balance of nature in the form of good wishes (*mangal kamana*), which means that speed of air may be beneficial, temperature of sun may be beneficial, rain from thundering clouds may be beneficial, day may be beneficial and night may be beneficial. Indra, Agni, Varun and other gods may be beneficial. Drinking water and rainwater may be beneficial. Earth may be good for our residence. Water may be pleasant to us and we may take it for getting power. We may win natural wars. *Dyau-Loka* (upper worlds) may be peaceful, space may be peaceful, earth may be peaceful, water may be peaceful, medicines may be peaceful, vegetation may be peaceful, all the forces of the world may be peaceful, wisdom may be peaceful, everything may be peaceful. Peace itself may be peaceful, and this peace may prevail throughout life (*Yajurveda* 17,36).

13

Hindu Dharma as Panacea for Environmental Crisis

Sanjay Kr. Shukla

Man has shown himself capable of the knowledge to give him certain mastery over environment. The danger to man in the future comes not from nature, but from man himself. John Passmore, in his *Man's Responsibility for Nature*, argues that the old testament (*Genesis* in particular) not only confers on man, dominion over nature, but also leaves open the possibility for an attitude of absolute despotism towards nature on the part of mankind.[1] Man must let go his technological and material narcissism because there cannot be real, responsible and effective environmental ethics in a world "dominated by technological mentality and crass materialism". Only by creating a workable environmental ethics and world conservation strategy can there be a chance for human survival. What is needed today is to remind ourselves that nature cannot be destroyed without ultimately mankind being destroyed itself. Centuries of rapacious exploitation of the environment have finally caught up with us, and a radically changed attitude towards nature is now not a question of spiritual merit but of sheer survival. The problem of environmental degradation and ecological imbalance has put forth the danger in which our human existence is at stake. The scientific and technological advancement and western religious beliefs and philosophical ideas are to a great extent responsible for putting us in such a perplexed situation. The religious beliefs of Semitic religions and scientific strides have provided enormous material comforts to human beings but we should not feel any hesitation in confessing that they have brought human existence on the brink of disaster. It is from historical point of view we find that man is fashioned according to his own faith and belief. The notion of alienation can be traced to the Bible where it is stated.. "Then God said let us make men in our own image, in our likeness, and let them rule over the fish or the sea and bird, or air or livestock, over all the earth and over all the creatures that move along the ground" (*Genesis* 1.26). This statement provides us license to exploit nature in every fashion. The historian Lynn White has criticized Judeo-Christian attitude towards the environment. He argues that the dualism of spirit vs. matter lies at the root of our contemporary environmental problems. Christian cosmology is based on the notion of creation *ex nihilio*, created out of nothing. This doctrine teaches that God created the whole cosmos with all animate and inanimate beings out of nothing. It secularizes nature and when nature is secularized, it becomes easy for human beings to objectify nature and control it. White accuses Christianity of being most anthropocentric (man centred) religion the world has ever seen. He believed that man has been given the dominion over all of God's creation and Christianity made it possible for man to exploit nature. Protestant theology interpreted the doctrine of creation in such a way as to establish an absolute distinction between God and creation. The dichotomy between God and the world logically implies secularization of nature and thereby rendering justificatory mechanism to dominate and exploit the nature by human beings. Contemporary theologians argue for stewardship in place of domination theme. "The Lord God took the man and put him in the Garden of Eden to work it and take care of it" (*Genesis* 2:15). Both these motifs in the final analysis objectify nature for anthropocentric ends. The only difference with the stewardship theme is that man is placed at the centre of nature, while with the domination theme man is above nature. Christian theologians are now willing to consider a unity theme as a basis of human world relationship.[2] Unlike the domination and stewardship themes, which survive on the distinction between human beings and nature, the unity theme affirms the co-substantiality of both human being and the world.

II

Protagoros upholds man as the measure of universe (*Homomensura*), which justifies man's liberty to freely exploit the nature. Franscis Bacon stressed the importance of the objects of sense perception and the mere passivity of the subject which culminated in the process of separating subject and object. Rene Descartes separated the world into the realms of rescogitans and resextensa which only added support to the developing belief in a natural world devoid of intrinsic worth.[3] Hence, from the seventeenth century onwards, empiricism, together with the separation of facts and values, became the intellectual orthodoxy in the development of modernity and in similar fashion rationalistic belief is responsible for separating spirit and mind from the objective world of nature. This trend culminated in the twentieth century with the thrust towards 'objective' forms of philosophy and social theory especially in analytic and linguistic philosophy. Max Scheler rightly pointed out that "to conceive the world as value free is a task which men set themselves on account of a value: the vital value of mastery and power over things".[4] With the writings of such scientists as Kepler and Galileo the development of dualistic thinking was further intensified—in metaphysics between mind and body, in cosmology between God and nature and in epistemology between rationalism and empiricism. On keeping with the image of a machine, the Newtonian model was essentially atomistic and reductionistic in outlook. It promised a worldview that can be comprehended in terms of so many particles of matter in motion. The new science promised the transformation of man himself into measurable and manipulative parts of the great machine. The man has disappeared as subject and reappeared in the world as object. Mind itself was dissolved into particles into motion by materializing solvents of the new physics.[5] It is true that quantum mechanism believes in the fundamental unity of the universe but even then scientific technology is providing instruments to us for conquering nature. The manner in which capitalist countries are propagating consumer culture will lead us to the state of *'practico-inertia'* where human consciousness instead of determining the significance of objects will be determined and directed by them. It is the intention to say that man being defined in terms of *'Homo-fabeur'* will become the product of its own product and lead alienated life from nature. Hence, the attempt of man to win over nature will be nothing else except sisiphian failure; ancient people have realized that destruction of plant will lead to the destruction of environment. *Mahabharat* is a war between the divine and the demon (Daivi Sampat and Asuri Sampat). Divine forces are unselfish and fostering equilibrium and harmony in the universe, while demonic forces are selfish and tending to spread disequilibria and disharmony in the universe. Man is called to develop the *daivisampat* which respects all existent beings alike and which advocates the utilization of resources of nature with a sense of gratitude and humility. The moto of our daily prayer is that let me coexist peacefully with all these beings and let me not sin against any one of these-

Ma Vighnam Ma Ca Me Papam Ma Ca Me Paripanthinah!
Dhruyo Jayo Me Nityam Syat Paratra Ca Shubgatih!!

Yojna is its cosmic setting and the concept of *loksangraha* also point to the same ideal. Preservation of ecological balance in nature is good for all beings. We must strive for the good of all and undertake activities that promote universal happiness (Dharma) and whatever hinders this is Adharma. The ecological awareness can be very beautifully located in *Manusmriti* as we learn that man should protect the moveable and immoveable components of environment by taking into consideration their usefulness for human health and welfare. The value of organic manure and humus in cultivation was known long before Manu's period as in *Taitriya Samhita* stated *Saktyadhana Sadhanam Gomyadi Dravyam Matyam*. We are now recognizing that the excessive use of chemical fertilizers and pesticide have disturbed the entire land, which grew out of religious feelings and divine sentiments. The earth is not simply a natural object but a loving mother who sustains all beings. He therefore prays to the mother earth every morning because he regards himself guilty of touching his mother with his feet.

Samudravasane Devi Pravatastanmandale!
Vishnupatni Namastubhyam Padsparsam Ksamasva Me!!

The Indian approach exhibiting relation between them recommends coherence with natural order by psychophysical transformation in human beings in place of producing changes in natural order and environment. Here, we find in place of man opposing nature and nature opposing mankind, ontic identity between them, the identity of microcosm and macrocosm and entire creation is being governed by extraordinary moral cosmic order (Rita). In Yajurveda it is prayed that sky, atmosphere, earth, water and plants be peaceful. May all the learned persons, God and Veda be peaceful? May peace itself be peaceful and let peace come unto me-

Om dyauh Shantih, Antariksah Shantih Prthvi Shantih, Apah Shantih, Oshadhyah Shantih, Brahm Shantih, Sarvam Shantih, Shantireva Shantih, Sa ma Shantiredhi (Yajurveda 36-17.)

Environment consists of biotic (living organism) and abiotic (nonliving materials) factors and plant is recognized as major biotic factor in Patanjali Mahabasya. The felling of tree is condemned in the verse "Atmapradha Vrksanam Phalanyetani Dehinam". The mechanical order has been associated with a framework of values based on power;

fully compatible with the directions taken by commercial capitalism.[6] The concept of progress represents a central value for the modern western world. No longer would people have to look towards some Augustinian city of God. Human salvation could be sought in secular terms. Hence, the idea of progress became intimately tied to the notion of 'want' that might one day be ended through the progressive subjugation of the natural world for human needs.[7] The exploitation of weak is an inherent part of the material culture, which believes in the philosophy of "survival of the fittest". The sole aim of life in this culture is to enjoy all the transitory sensual pleasure of the world and is the root cause of greed, misery and disharmony in the world. Hinduism, on the other hand, believes in the doctrine of *Jeevastha Jivyasam*—live and let live and *Kamaye dukha taptanam praninam artinashanam*—the air of life is to remove the suffering of living beings.

III

It is obvious that the Hindu Rishis and the Upanishadic era perceived harmonious relationship between the needs of man and spectacular diversity of the universe. They did not subscribe to the prevailing western worldview that true nature of man was essentially to dominate and control nature by all possible means. It is noteworthy that man in Vedic age treats himself as a part of nature. It is said that the whole world is a creation of sacrifice undergone by cosmic Purusa. It may be further noted that the relation of man with rivers in Vedic time was that of a worshipper and the worshipped.

It is not only capitalist ideology, which is responsible for ecological crisis, but socialism is no less guilty for it. Karl Marx argued that with the end of the class system, humans would at last be able to wage war against nature. Indeed both capitalist and socialist ideologies have bowed before the expansionist ethos and pitted man and his technology against the natural world. Marx in the Grundrisse praises the industrial revolution brought about by capitalism in as much as it was giving human kind the necessary mastery over nature. Both Ophuls and Heilbroner argue that the values underlying capitalism and socialism are predicated upon the transgression of the bounds and limits of nature.[8] Ehrenfield refers to this promethean like mentality of the contemporary industrial world as the "arrogance of humanism—the hubris for which we are now having to pay the price".[9] The history of the West is the progressive removal of mind or spirit from phenomenal appearances. The phrase '*die Entzauberung der Welt*' (disenchantment of the world) of Weber and '*die Entgotterung der Natur*' (the disgodding of nature) of Schiller points to human psyche and desire to subjugate the nature. The whole materialist position assumes the existence of the world "out these" independent of human thought. One thing that is certain about the history of western consciousness is that the world has been progressively disenchanted or disgodded.

Caroline Merchant contends that the development of modernity went hand in hand with the de-animization and hence 'objectification' of nature. As a consequence, nature was no longer regarded as a 'thou' but as an 'it'—to be exploited for utilitarian ecological balance. Modern earth is becoming more and more saline and thereby unproductive in many areas. The *Manusmrti* distinctly states that the trees on account of their being under the influence of tamoguna possess a sort of dormant or latent consciousness but are capable of pleasure and pain—*Antah Samjna Bhavanti Iti Sukh dukh Samanvitah* (*Manusmrti* 1-49). World community now accepts this belief after experiments made by J.C. Bose. In *Manusmrti*, it is stated "one should not throw urine or fasces into the water, nor saliva, nor blood, nor poisonous beings (*Manusmrti* 4-56) and which clearly proves serious concern against water pollution. *Jivo Jivasya Jivanam* could be interpreted to mean that one form of life promotes the welfare of another life without consuming or damaging it. Hence, holistic Hindu view of life and entire Indian philosophical tradition is the right way of establishing relationship between man and nature.

IV

The principle of the sanctity of life is clearly ingrained in the Hindu religion. Only God has absolute sovereignty over all creatures including man and therefore man cannot act as a viceroy of God over the planet, nor assign degrees of relative worth to other species. The religion provides no sense of absolute superiority of man over nature. In *Atharveda* it is clearly stated that earth is not for human beings alone, but for other creatures as well. It is considered in Hindu belief that all that exists has been created by the Supreme Being, comes from the Supreme Being, and will return to the Supreme Being, which is an adequate basis for the veneration of the natural world in which man finds himself.[10] The most important aspect of Hindu theology is the association accorded to different species with reincarnation and deities. It is believed that Supreme Being has actually got himself incarnated in the forms of various species. The justification for animal worship is to be found in the incarnation doctrine. In *Yajurveda* (13.47), we find that no person should kill animals which are helpful to all, and by serving them one should obtain heaven. He who inures innocuous being with a desire to give himself pleasure, never finds happiness, neither living nor dead. (*Manusmrti* 5.45). In Hindu religion, nature has been regarded as indispensable in the life of human beings. It has also been considered as revered, bestowal of good, and protector from the evils with a concept of God living in it. The sense of worship that developed among the Hindus has a background of utility and spiritfulness. Hindu society has been much aware of the fact that flora of the country has a protective power and indiscriminate destruction would result in pollution and chaos in the society. Some of the plants and trees have been well known for their medicinal power, for their spiritual power, and some as an abode of God. Hence, for this reason in the times of *Rigveda*, tree worship has been quite popular and universal. In *Padma Purana* (56.40-41) it is stated that cutting of trees is a punishable offence and the person who indulged in cutting of trees and destroying the grass field had to go to hell.

Hindu culture does not accord a superior status to human beings over their natural surroundings. Hence, their traditional attitudes towards the ecosystem (e.g., animals and birds, plants and trees) has been kind and respectful. Charak had written extensively about *vikrti* (pollution) and warned about the side effects of seasons, foul air and polluted water. Hindus consider water as a powerful media of purification and also as a source of energy. The water in the sky, the water of rivers, water in the well, whose source is ocean, may all these sacred waters protect me (*Rigveda* 7.42.2). The healing property and medicinal value of water has been universally accepted provided it is pure and free from all pollution. Charak, while talking about *vikrti*, mentioned air pollution especially as a cause of so many diseases. The concern for environment can be gleaned from this passage: "A person who is engaged in killing creatures, polluting wells, and ponds, and tanks, and destroying gardens, certainly goes to hell." (*Padma Purana, Bhoomikhanda* 96.7-8). Hinduism does not subscribe to the western worldview that man's role on the earth was to exploit nature for his selfish purpose. Man in Hindu culture, was instructed to maintain harmony with nature and to show reverence for the presence of divinity in nature. Hence, Hindu religion (including other religions of India) can be of a great help in transforming society from its current preoccupation with materialism and consumerism to a conserver society.

Notes

1. John Passmore, *Man's Responsibility for Nature*, pp. 26-27.

2. G. Barbour, Ian, *Technology, Environment and Human Values,* pp. 14-15.
3. Descartes, Rene, "Animals are Machines," in T. Regan (ed.), *Animal Rights and Human Obligation,* pp. 60-66.
4. Skolimowsky, H., *Eco-philosophy,* p. 16.
5. Floyd W. Maston, *The Broken Image,* p. 13.
6. Caroline Merchant, *The Death of Nature: Women, Ecology and the Scientific Revolution,* p. 193.
7. J. Bury, *The Idea of Progress,* p. 73.
8. Ophlus and R. Heilbroner, *Ecology and the Politics of Scarcity: An Enquiry into the Human Prospect.*
9. D. Ehrenfeld, *The Arrogance of Humanism.*
10. K.W. Morgan, *The Religion of the Hindus,* p. 102.

14

Hinduism and Environment

Shiv Bhanu Singh

The twentieth century is well known for revolutions in the field of technology and development of splendid technological means for bringing about spectacular changes in our living environment. On the other hand in the same century, humanity has shown its utmost concern for environment. "nurture nature", "back to nature", "eco-friendliness" are the phrases which are famous in our contemporary age. These phrases seem to be the direct reflections of our candid confessions of eco-insensitivity. It is unfortunate that the most intelligent species on the earth live such an eco-unfriendly life on it. It is surely a glaring example of human folly.

The ecological imbalance and environmental crisis are the result of western attitude towards nature. According to the dominant western tradition, the natural world exists for the benefit of human beings. God does not care how we treat it. Human beings are the only morally important members of this world. Nature itself is of no intrinsic value and the destruction of plants and animals cannot be sinful unless by this destruction we harm human beings.

In the later half of the twentieth century a changed attitudes towards nature evolved and reverence for life was advised. More than forty years ago the American ecologist Aldo Leopold wrote that there was a need for a 'new ethic'—an ethic dealing with man's relation to land and to animals and plants which grow upon it. His proposed 'land ethic' would enlarge the boundaries of the community to include soils, water, plants and animals, or collectively—the land. Another Norwegian philosopher, Arne Naess wrote a brief but influential article distinguishing between 'shallow' and 'deep' stands in the ecological movement. Shallow ecological thinking was limited to the traditional moral framework. Deep ecologists, on the other hand, wanted to preserve the integrity of the biosphere for its own sake, irrespective of the possible benefits to humans that might flow from so doing. In a paper published in 1984, Arne Naess and George Sessions, an American philosopher involved in the deep ecology movement, set out several principles for a deep ecological ethic, beginning with the following:

1. The well-being and flourishing of human and non-human life on earth have a value in themselves (intrinsic value). These values are independent of the usefulness of the non-human world for human purposes.

2. Richness and diversity of life-forms contribute to the realization of these values and are also values in themselves.

3. Humans have no right to reduce this richness and diversity except to satisfy vital needs. Two Austrians working at the deep end of environmental ethic, Richard Sylvan and van Plumwood also extend their ethics beyond living things, including in it an obligation not to jeopardize the well-being of natural objects or system without good reason. In *Deep Ecology*, Bill Daval and George Sessions defend a form of 'bio-centric egalitarianism'. The intuition of bio-centric equality is that all things in the biosphere have an equal right to live and blossom and to reach their own individual form of unfolding and self-realization within the larger self-realization. This basic intuition is that all organisms and entities in the ecosphere, as parts of the interrelated whole have equal intrinsic worth.

Reverence for nature, the principle of the sanctity of life, bio-centric equality and cosmo-centric attitude can be very easily found in Hindu religion ever since ancient times. Hindu religion believes that only God has absolute sovereignty over all creatures including man. Thus, man had no dominion over his own life or non-human life. Hindu religion does not provide any sense of absolute superiority of man over nature. The idea of the Divine Being as the one underlying power of unity is beautifully expressed in *Yajurveda*. The loving sage beholds that Being, hidden in mystery, wherein the universe comes to have one home; Therein unites and therefrom emanates the whole. The omnipresent pervades souls and master, like warp and woof in created beings (*Yajurveda* 32.8)

The sacredness of God's creation means, no damage could be incurred on other species without adequate justification. Therefore, all lives, human and non-human, are of equal value and all have the same right to existence. According to *Atharvaveda,* the earth is not for human beings alone but for other creatures as well.

In Hindu religion, we find proper respect and consideration of the natural word. This includes the flora and fauna of the earth, and creatures in the sky and under the sea. It is considered in Hindu belief that the Supreme Being has created all that exists, comes from the Supreme Being, and will return to the Supreme Being. It is an adequate basis for the veneration of the natural world in which man finds himself. The history of human survival stands as a good example of the interdependence, cooperative living and close association, which exist, among the forces of human and non-human nature. The Hindu concept of the divinity of nature, somewhat reminiscent of a form of nature worship, came from an earlier age. At one place in *Srimad Bhagavata* (7,14,9) we find that it is advised by the seers to treat all other species like their own children. One should look upon deers, camels, monkeys, donkeys, rats, reptiles, birds and flies as though they are their own children. We find a large number of incarnations of God in Hindu religion. God incarnated himself in the form of fish, tortoise, boar and dwarf. As Rama, he was closely associated with monkeys and as Krishna, he was always surrounded by the cows. Monkeys received the status of sacred animals because one of their ancestors *Hanuman* rendered an undying service to Rama in his great adventure to Lanka. Several Hindu gods and goddesses have animals and birds as their mounts. Thus we find that animals and birds are not only given respect but also receive reverence in the Hindu society, these two factors have provided a solid foundation for the doctrine of Ahimsa, non-violence against animals and human beings alike. In *Rigveda* (10.87.16) and *Yajurveda* (13.47) animals are prevented from being killed. In *Manusmriti* (5.45) *Narsinhapurana* (14.44) *Visnupurana* (3.8.15), the concept of non-violence has been propounded.

By the end of the Vedic and Upanisadic period, Buddhism and Jainism came into existence, and the various kings practicing these religions further strengthened the protection of animals, birds and vegetation. The Buddhist emperor Ashoka (273-236 B.C.) undertook various steps to safeguard the environment. He promoted, through public proclamation, the planting and preservation of flora and fauna. Pillar edicts, erected at various public places, expressed his concern about the welfare of creatures, plants and trees and prescribed various punishments for the killing of animals, including ants, squirrels and rats.

In Hindu religion, nature has been regarded as indispensable in the life of human beings. It has also been considered as revered, bestower of good and protector from the evils, with a concept of God living in it. The sense of worship that developed among the Hindus has a background of utility and spiritfulness. Hindus consider trees and plants as the abode of various gods and goddesses. Hindu society has been much aware of the fact that the flora in the country has a protective power and indiscriminate destruction would result in pollution and chaos in the society. Some of the plants and trees like Tulsi, Pipal, Vata, Bela, Ashoka, Amalaki, Neem, etc., have been well known for their medicinal power, spiritual power and some as abode of God. As early as in the times of *Rigveda*, tree worship has been quite popular and universal. Cutting of trees and destruction of flora were considered as sinful act. In *Padma Purana* and *Charak Samhita*, cutting of trees and destruction of forests were considered as punishable offence and dangerous act for humanity.

Thus, it is clear that Hindu culture does not accord superior status to human beings over their natural surroundings. Hence, their traditional attitude towards the ecosystem (e.g., animals and birds, plants and trees) has been kind and respectful. Maintaining proper sanitation was considered to be the duty of every one. When Charak wrote about Vikrti (pollution) he wanted people to be warned about the side effects of seasons, foul air, and polluted water. He further mentioned air pollution as cause of so many diseases. Hindus were also cautious about those activities, which were detrimental in any way to the quality of environment. In Hindu culture, man was instructed to maintain harmony with nature and to show reverence for the presence of divinity in nature. Hindu culture in ancient and medieval times provided a system of moral guidelines towards environmental preservation and conservation. Environmental ethics was practiced not only by common man, but even by rulers and kings. Conclusively, we can say that Hindu religion can be of great help in transforming society from its current preoccupation with materialism and consumerism. A moral awakening can strengthen various laws enacted to conserve society for environmental protection. Religion can exert a unique moral leadership among its followers, particularly with respect to strengthening man's harmony with nature so that environmental crisis may be overcome.

References

Arnold, J. Toynbee (1996), *Change and Habit*, Oxford University Press, London.

Dwivedi, O.P. and Tiwari, B.N. (1987), *Environmental Crisis and Hindu Religion*, Gitanjali Publishing House, New Delhi.

Leonard, J. Weber (1976), *Who Shall Live?*, Populist Press, New York, pp. 41-42.

Morgan, R.W. (1943), *The Religion of the Hindus*, Ronald, New York, p. 102.

Thomas, S. Derr (1975), "Religion's Responsibility for the Ecological Crisis", *World View*, 18 p. 43.

15

Nature, Environment and Spirituality

Rajjan Kumar

Eco-centric reflections are vital and crucial contemporary issues. Philosophers have played an important role in this direction, as the philosophical reflection has done a great deal to clarify and illuminate useful strategies for dealing with our global ecological crises.

Simple Value-Rational Questions and Environment

Eco-centric debate enjoins several issues, queries, questions and most of them are value-rational. As our knowledge of environmental devastation and the global consequences of this devastation grow, so do questions about how we ought to think about and act towards the natural world. The progress of science blurred the human-non-human boundary like human interest with nature. It is environmental ethics, which challenges the separation of science and ethics, trying to reform a science that finds nature value free and an ethics that assumes that only humans count morally. It has to evaluate nature, both wild nature and the nature that mixed with culture and to judge duty thereby.

We live in an unprecedented historical period. Within the past few decades the actions of mankind have become, for the first time ever, a threat to life as we know it; not only through nuclear disasters that may or may not happen but through changes in the global ecology that are happening. Earlier, the continuance of life on a planet was taken as given rather than subject to a question mark. No wonder, then, it has been said we live in a post-era – post-rational, post-enlightenment, post-modern, post-foundationalist, post-structuralist, etc., If anyone phenomenon distinguishes the start of a new era and a post-condition, it is this: humanity's newly achieved ability to effectively destroy its own sustenance. The world has become post-immoral; not in the sense that life on the planet is necessarily moral but rather that there is no longer any assurance of its immortality. We live in a world-at-risk, where life has become contingent upon our own actions.[1]

Ethics, the branch of philosophy, seeks an appropriate respect for life, and environmental ethics no longer away from this content. It is true that the environmental ethics stretches classical ethics to the breaking point. But we do not need just a humanistic ethic applied to the environment, as we have needed one for business, law, medicine, technology, international development, or nuclear disarmament. Respect for life does demand an ethic concerned about human welfare, an ethic like the others and now applied to the environment. But environmental ethics in a deeper sense stand on a frontier, as radically theoretical as they are applied. They alone ask whether there can be non-human objects of duty.[2]

Environmental ethics require risk. They explore poorly charted terrain, where one can easily get lost. One must hazard the kind of insight that first looks like foolishness. Some people approach environmental ethics with a smile expecting chicken liberation and rights for rocks, misplaced concern for chipmunks and daisies. Elsewhere, they think, ethicists deal with sober concerns: medical ethics, business ethics, justice in public affairs, questions of life and death and of peace and war. But the questions here are no less serious. The degradation of the environment poses as great a threat to life as nuclear war, and a more probable tragedy.[3]

Ecosystem and Environmental Ethics

No environmental ethics has found its way on earth until it finds an ethic for the biotic communities in which all destinies are entwined. "A thing is right," urged Aldo Leopold. When it tends stop preserve the integrity, stability, and beauty of the biotic community. It is wrong when it tends otherwise."[4] Again, we have two parts to the ethic: first, that ecosystems exist, both in the wild and in support of culture; second, that ecosystems ought to exist, both for what they are in themselves and as modified by culture. Again, we must move with care from the biological assertions to the ethical assertions.

Classical, humanistic ethics find ecosystems to be unfamiliar territory. It is difficult to get the biology right and, superimposed on the biology, to get the ethics right; fortunately, it is often evident that human welfare depends on ecosystemic support, and in this sense all our legislation about clean air, clean water, soil conservation, national and state forest policies, pollution controls, renewable resources, and so forth is concerned about ecosystem-level processes. Furthermore, humans find much of value in preserving wild ecosystems, and our wilderness and park system is impressive.

Still, a comprehensive environmental ethics needs the best, naturalistic reasons, as well as the good, humanistic ones, for respecting ecosystems. Ecosystems generate and support life, keep selection pressures high, enrich situated fitness, and allow congruent kinds to evolve in their places with sufficient containment. The ecologist finds that ecosystems are objectively satisfactory communities in the sense that organismic needs are sufficiently met for species to survive and flourish, and the critical ethicist finds (in a subjective judgement matching the objective process) that such ecosystems are satisfactory communities to which to attach duty. Our concern must be for the fundamental unit of survival.

The plants and animals within an ecosystem have needs, but their interplay can seem simply a matter of distribution and abundance, birth rates and death rates, population densities, porosities, and predation, dispersion, checks and balances, and stochastic process. An ecosystem is a productive, projective system. Organisms defend only their selves, with individuals defending their continuing survival and with species increasing the numbers of kinds.

Contrary to this, the evolutionary ecosystem spins a bigger story, limiting each kind, locking it into the welfare of others, promoting new arrivals, increasing kinds and the integration of kinds. Species increase their kind, but ecosystems increase kinds, superposing the latter increase onto the former. Ecosystems are selective systems, as surely as organisms are selective systems. The natural selection comes out of the system and is imposed on the individual. The individual is programmed to make more of its kind, but more is going on systemically than that; the system is making more kinds.

Environmental ethics, in the humanistic sense, will say that ecosystems are of value only because they contribute to human experience. Humans count enough to have the right to flourish in ecosystems, but not so much that they have the right to degrade or shut down ecosystems, not at least without a burden of proof that there is an overriding cultural gain. In a comprehensive ethics of respect for life, we ought to set ethics at the level of ecosystems alongside classical human-istic ethics.

Nature as Spiritual Guide and Teacher

Spirituality is the fundamental concern of Indian thinkers and environ-mental ethics has not been an exception of the very nature of Indian culture. Investigation of nature has been a major concern in Indian tradition. Anyone familiar with the epic and dramatic literature of India will recall numerous vivid and loving descriptions of nature offered therein. They all define different themes in several ways, however, they do not ignore the aptitude of environmental ethics.

The beauty of the forest through which Rama and Sita proceeded, the peace and tranquillity of the famous hermitages visited by the Pandavas, the greatness of the Himalayas, and the awesomeness of the Vindhyas are immortalized in a beautifully articulated language. It has remarked on the richness of vocabulary employed in these descriptions of nature, the astonishingly great number of species of fauna and flora identified by name and described in detail. Surely, Indian thinkers have not been unaware of the beauty and greatness of nature—a nature so luxuriant in many parts of the country as to be almost without comparison.[5] This variegated nature serves as the background of human spirituality and as a scaffolding for divine intervention in environment ethics.

Over and against this, the way in which nature is seen as a spiritual guide to humanity, as related in the *Bhagvata Purana*. A young ascetic, identified in another passage as Dattatreya, relates how he had adopted nature as guide to wisdom and liberating knowledge.[6] The account of the teachings of the "twenty-four guru(s)" is prefaced by words put into the mouth of the Lord: The investigators of the true nature of the world are uplifted by their own efforts in this world. The self is the infallible guide of the self: through direct perception and through analogy one can work out one's salvation.[7] It is suggested that "true knowledge of nature" leads to "true knowledge" of self and God.

The twenty-four guru(s) from nature, which Dattatreya has chosen to follow, induce him to adopt practices and rules for his life, which reaffirm his ideal *Samnyasa*, and through it contribute to his liberation. Nature acts precisely as the human guru does: proposing through words and by example a path leading to insight and realization. A sampling from the lengthy text will suffice to make the point, which directly commence with the fact of environmental ethics.

The earth has taught Dattatreya steadfastness and the wisdom to realize that all things, while pursuing their own activities, do nothing but follow the divine laws, which are universally established. Furthermore, the earth has taught him that existence in a body is a being-for-others (*parartha*) to be lived out in humility and forbearance. Fire, too, is an excellent teacher and an example for the ascetic being "full of splendour and made brighter by the glow of *tapas*. Not sullied by what is consumed …Sometimes hidden, sometimes visible, assuming the shape of the fuel which it consumes, burning up past and future sin".[8] The honeybee teaches the student to go out and collect the essence from all possible entities. It also provides a negative lesson. Do not hoard any food. To substantiate this part of the bee's teachings, the text recounts a popular story about a bee that perished together with its stored-up supply of food.[9]

Finally, Dattatreya learns the most decisive lesson from his own body: "This body, subject to birth and death and constantly and ultimately a source of affliction, is my guru as it prompts me to renunciation and discrimination. Though it helps me to contemplate, it really belongs to others. Realizing this I am going forth, renouncing all." And it defines the spiritual nature of environmental ethic, which needs the best, naturalistic reasons, as well as the good, humanistic ones, for respecting ecosystems because ecosystems generate and support life.

Consequently, the Indian spiritual environmental ethicist, Dattatreya through physics, biology, anthropology, and psychology has reached a stage of wisdom, which makes him aware of "the true nature of things" and delivers him from any need to transform nature into consumer goods.

Nature, Spirit and Environmental Ethics

Inquiry into the nature of 'nature' is most debating issue, as it does not aim at appropriating nature through the senses, or their extension in technology, and it does not cultivate instrumental reasons. Now, the question is raised as to what is the appropriate sense of inquiring into the nature of nature? Samkhya, the most prominent school of India, who is actively engaged in disclosing the fact of nature, has promptly explained that the proper attitude for this science of nature is detachment from the sense appetites, so as to let the subjectivity of nature appear as it is, before it has been distorted by human interference. This science aims at the knowledge of nature, not at its use.

In its most generic sense, 'nature', understood as *prakriti*, is not seen as 'substance' to which certain qualities are added; rather, it is defined as equilibrium of three *guna(s)* without particular name or form. The very choice of the 'qualities' universally identified with nature and their applicability to human nature, seem to be further proof that Samkhya deals with the subjectivity of nature rather than its objectivity.

Furthermore, it is considered that this is what distinguishes the modern scientific inquiry into the nature of 'nature' from various levels of thoughts. Vedanta has a very different kind of orientation of thinking along with different interest. Vedanta is interested in finding out about what nature is in relation to consciousness, whereas modern science wants to find out how it works. The vedantic tradition, at whose core is *vivek* (differentiation) and which at the same time insists on the oneness of reality, represents a path of liberation through nature, a fulfilment of the human person's own destiny worked out not over and against the exploration of nature but through it.

The experiments, which Svetaketu is asked to perform by his father Uttalaka Aruni, dividing a fig, dividing the seeds, mixing salt and water, tasting the mix, quite clearly combine instrumental and reflective analysis of nature. From the question—what do you see? the questioner leads to the problem of where does it come from? And what is it? From an examination of what appears to be part of the external world the question leads to an investigation of the self.[10]

Meanwhile, we have learned to see a great deal inside the seed of the fig; where Svetaketu could see nothing at all, we see the fine material structure of the cell, the zygotes, the molecules of the amazingly numerous substances, etc. And what Uttalaka Aruni considered the atom, has become the subject of big science research, which finds ever-smaller subatomic particles.

To consider more, we think of more philosophy. Maya, one of the most considerable facts of Indian philosophy, contains *vasana* (impressions, potential developments, natural laws) of the entire universe.[11] Consciousness reflects (actively) in these states (namely, waking and dreaming) the mental imprints (*buddhivasana*). The seed of the world in association with this reflection emerges in the form of intellect (*dhi*). In the mind (*buddhau*), the consciousness reflection (*cidabhasa*) is unclearly reflected. This of things, nature being grounded and contained in an entity both neither scientifically nor logically ascertainable nor manageable, should not create the impression that we are talking about an 'illusion' in the everyday world.

It is further stated that nobody has the power to alter the world of waking and dream states.[12] Individual persons as well as the Lord (God) himself are "reflections (abhasa) in maya of atmah". Its reality obviously is accessible only through a third path, a sensibility not employed in science. Between Isvara and Maya there is a kind of mutuality (both being ultimately *acintya* (logic). Isvara is the maya-reflection of consciousness and at the same time its inner ruler (*antaryamin*), omniscient (*sarvajna*), and womb of the world (*jagadyoni*).[13]

The Lord, as the "bliss-sheath", is the carrier of all *vasana(s)* (information potentials) of all living beings[14] that is his omniscience. As all pervading, he is also "all-supporting (Sarvopadana, material cause of the universe).[15] He is detectable as 'Inner Ruler' through an analysis of the fine structure of the universe, where the progress from the subtle to the subtler stops. This Lord is the source of the universe (*jagadyoni*), in so far as to him is due the 'manifestation' (*avirbhava*) and 'demanifestation' (*tirobhava*) of the world.

Now, we see the fact, that has been propagated by Indian seers in a philosophic manner, possesses great information for environment ethicists. Environmental ethics is eco-centric, which finds that ecosystem is a productive and projective system. The system is a game with loaded dice, but the loading is a profile tendency, not a mere stochastic process. An ecosystem has no head, but it heads towards species diversification, support and richness. Though not a super organism, it is a kind of a vital field. In a comprehensive ethics of respect for life, we ought to set ethics at the level of ecosystems alongside classical, humanistic ethics.

In practice, the ultimate challenge of environmental ethics is the conservation of life on earth. In environmental ethics, one's beliefs about nature, which are based upon (but exceed) science, have everything to do with beliefs about duty. The way the world is, informs the way it ought to be. We always shape our values in significant measure in accordance with our notion of the kind of universe that we live in, and this process drives our sense of duty. It can be concluded as—"Taddure Tadantike".

Environment, Spirituality and Jainism

'Environmental ethics' as a nomenclature of concept although is new and contemporary, but is not new for Jainas, the live and leading representative of Sramanica tradition, the one aspect of the composite culture of India. Another part is called the Vedic tradition. Environmental enlightenment is an indigenous temptation of Indian culture, accordingly

Jainism has indifferent aptitude. Jainism, a schiasm preached by Jinas possess religious, philosophical discourses, amazingly it considers equal attentions for environment and environmental ethics, which of course is eco-centric but is forwarded in the spiritual manner.

As it is well proved that Jainism has given much attention towards animism, *ahimsa* (non-violence) plays the key-role to open the different viewpoints of Jaina spiritualistic ideologies of environmental ethics. It is well known that non-violence, in the sense of regard for life (as environmental ethicists believed) is accepted as an ideal by almost all the religions of the world, but none pursue it as meticulously as Jainism, which prohibits not only the killing of human beings and animals (having five senses with mind) but even of plants and alike organisms (having one sense only). To hurt one-sense living beings like plants, etc., is also considered to be an act of violence or *ahimsa*.

Earth, water, fire, air and plants—all five are considered as animated entities having one sense that is touch. Respect or regard for life should be total as well as unconditional. Life, in whatever form it may be, should be respected; we have no right to deprive a living being of its life. The text *Dasavaikalika* says: "Every living being wants to live and not to die; this is the reason why the sages prohibit violence."[16] Although Jainism sets its goal to the ideal of total ahimsa, external as well as internal, yet the realization of this ideal in practice is by no means easy. Non-violence is a spiritual ideal fully realizable only on the spiritual plane and the concept of environmental ethics in no way differs.

A human being, however, is not just a spiritual but also a physical being and so cannot avoid all possible forms of injury either to himself or herself or his or her fellow creatures. Man can only strive toward a fully non-violent life, and only to the extent that he succeeds can he rise above the physical level. However, violence that is intentional can be avoided. Deliberate violence relates to our mental proclivities. A human is master of his or her thoughts, and so it is obligatory for all to be non-violent in this sphere. Other forms of violence are inevitable—namely, that violence necessary for self-defense and the execution of daily tasks—because a human, despite his or her inherently spiritual nature, lives also on the physical level.

At the physical level, human endeavours deviate from the path of spiritual order and needs arise to send man into his own natural path. Jainas put in many ways, but the most significant among all of them is *Karmadana*, which is directly associated with environmental ethics.

Karmadana is technicus terminus and is known as transgression (*aticara*) of *Upabhogaparibhoga Vrata*.[17] Its vow includes under the sub-class of right conduct as *Guhavrata*[18] (qualifying vow), which help to preserve and develop the five *anuvratas*. Regarding *upabhoga-paribhoga*, it is a vow which set limit to one's consuming and enjoying the things; so as to bring simplicity in life and also maintain the balance of ecosystem.

It helps to lead a moderate life by placing a limit on different objects. It grants a limited number of ways of earning and consumption of limited number of objects for one's living. *Karmadana* is directly related to livelihood and accordingly associated with environmental ethics, because its nature is completely eco-centric. *Karmadana* means putting a ban on the 15 cruel trades[19]—Angara karma (charcoal), Vana karma (wood), Sakata karma (carts), Bhataka karma (transport fees), Sphota karma (hewing), Darita vanijya (teeth), Taksakarma (lac), Rasavanijya (alcohol), Kesa vanijya (hair), Visa vanijya (poisonous articles), Yantra pinda (involving hilling), Nirtanchana (mutilation), Davagni dana (setting fire to a forest), Sarahsosana (drawing off water from lakes), Asatiposana (rearing anti-social elements).

The Jaina spiritual environmental ethic can be expressed— *Icchabahubina loye—a samjame niyatinca samjame ya pavattanani.*

Notes

1. *Applied Ethics: A Reader*, (ed.) E.R. Winkler and J.R. Coombs, "Aristotle, Foucault and Progressive Phronesis; Bent Flyvbjerg", p.12.

2. Applied Ethics: A Reader, "Environmental Ethics ... Holmes Rolston III", pp. 271-72.

3. Ibid., p. 272.

4. *A Sand County Almanac*, Aldo Leopold, p. 262.

5. *Hindu Spirituality: Vedas Through Vedanta*, (ed.) Krishna Sivaraman, "Spirituality and Nature", Klaus K. Klostermaier, p. 319.

6. *Bhagavata Purana*, 11.7.9; 2.7.4.

7. Ibid., 11.7.196.

8. Ibid., 11.9.2766.

9. *Hindu Spirituality*, "Spirituality and Nature", p. 320.

10. "Sa Ya Esoanima Aitad Atmyam Idani Sarvam Tal Satyam Sa Atma Tal Tvam Asi Svetaketo", *Chandogya Upanisad*, 6.13.3.

11. *Pancadasi* (Vidyaranya Muni), (ed.) Narayana Rama Acarya, 6.152.

12. Ibid., 6.160.
13. *V. Raghavan Felicitation Volume*, "The Advaitic Concept of Abhasa", Dr. Satya Deva Mitra, pp. 267-89.
14. *Pancadasi*, 6.161.
15. Ibid., 6.165.
16. Savve Jiva Vi Icchanti Jivium Na Marijjaum/Dasavaikalika, 6/11.
17. *Uvasagadaso*, 1/51.
18. *Tattavarthasutra*, 7/16.
19. *Uvasagadaso*, 1/51, *Sagaradharmamrta*, 5/21-23.

16

Buddhist Environmentalism

Avinash Kumar Srivastava

The story of present eco-crisis is a long history of man's march from the lush green forest to the juiceless dense jungle of the concrete. The history of human development in material terms is an interesting and intriguing story of human relationship with nature and natural phenomena. In fact, "with the emergence of *homo sapiens* on the planet earth the human history began, struggling with the problems of existence and survival like other living organism".[1] Nature sometimes nurtured and caressed him with her soothing weather and abundant supply of health giving food and drink, but sometimes threatened his existence by devastating forces like thunder, cyclones, quakes, etc. Being horrified by her scourging aspect, man primarily recoiled with anguish and surrendered before her considering it as a 'mysterious trauma'. This was man's initial defeat. But this helped develop man's ingenuity. The insecure feeling and sense of unprotection from natural calamities and wild animals kept this pursuit advance further. This generated in man an intense desire to distance nature as well as to create his own culture and climate. And then began man's bold and triumphant march against ravaging forces of nature. With gradual accumulation of knowledge and development of skill, man moved into the open fields, clearing the forest, tilling and ploughing the field. He ignited the kitchen hearth and started fortifying himself against the devastating forces of nature. In this process, he gained some knowledge of nature through some sort of naïve guesswork. "His will to exist naturally gave rise to will to know and his will to know was followed by his will to create....As natural consequence, his will to know the nature was the emergence of science.[2] This entire exercise was meant to free man from the clutches and control of nature as well as to defeat, dominate and distance her. With the rise of physical science dissection of nature for the purpose of examining and reassembling it according to human specification began. With the Western Enlightenment, development process increased rapidly. This led directly to industrial revolution, which not only changed the hue and colour of horizon but also the mind; leaping, hissing desires to possess and accumulate more and more riches led the industrialists to unethical, immoral and oppressive activities, and for the poverty stricken hungry mass, petty doles became the prime agents of systematic exploitation of nature. The invention of atom and hydrogen bombs caused irreparable damage to our ecosystem. Constant emission of poisonous gasses through chimneys and laboratories has verily resulted in the present ecological imbalance and pollution.

The urge to search ways and means to overcome the pollution problem gave birth to modern environmental movements and nature conservation programmes. There have been three international summits to tackle the critical situation, the first at Stockholm, second at Rio and the third at Johannesburg, but not much headway has been made. The perceptions differ between developed, developing and underdeveloped countries and any attempt to arrive at a common programme of action is bogged down. Recently, America refused to sign the UN resolution for an agreement restricting the emission of carbon and other harmful gasses beyond a certain limit because it will affect American economy adversely, and America is the greatest carbon producing country. The crux of the situation is that we are moving towards annihilation and if immediate efforts are not made honestly and seriously, doomsday will not await the change of our perception.

But for the vision of doom and gloom and their appropriate solution, Buddhists need not go to modern prophets of environmentalism. They have simply to give a repeated reading of their own texts. I am sure, that the proper interpretation of Buddha *Vacana*, in the present context, will give birth to Buddhist environmentalism, containing reason and remedy of impending tragedy. It is evident from Pali texts that Buddha, after apprehending eco-crisis, has warned against improper use of nature. He always advocated the proper management of natural resources. At the same time, he emphasized on the protection of nature from human encroachment. And thus stems out two environmental theories: (i) stewardship of nature, and (ii) protection of nature. A brief explanation of these two is needed here.

Stewardship of Nature

The prime purpose of this theory is to conserve nature in order to maintain the ecological balance as well as proper management of the natural resources, so that nature may not be misutilized and its resources are not wasted. Hence it presupposes two types of action plan: (1) Conservation of nature, and (2) Nature management.

Conservation of Nature

Nature and natural environment have played an important role in Buddha's life. He was born, achieved Bodhi[3] and attained Mahaparinibbana (death) at the root of trees.[4] He was brought up in a nature surrounding.[5] During his hard endeavour (from *Mahabhinishkramana* to the state of *Tathagata),* he spent his life in forests.[6] He gave his first sermon in the *Deer Park at Sarnath*[7] and his entire outdoor preaching were conducted under the shadow of trees. Buddhist monasteries were constructed in the natural surroundings and when dissention arose (for a short period) among monks he spend his time in the forest with wild animals.[8] He considered forest an ideal place for meditation practices.[9] He has recommended in his *Dhutanga Niddesa,* to live in the forest *(Arannikanga)*[10] and to live at the root of a tree *(Rukkhamulikanga)*[11] to the monks for their mental purification.[12] Hence nature and natural environment are intertwined in the "vain and vessels" of Buddha and Buddhism. That is why love for nature and its conservation is predominantly reflected in the Buddhist way of life. Even now the Buddhists revere old and large trees. Construction of parks and pleasure grooves for public uses are considered as auspicious work that fetches spiritual merits.[13] The Buddhist community all over the world has extended popular ritual of Bodhi-pooja, which is performed by pouring water to the root of Bodhi tree and paying respect to it, and other species of the tree. This ritual has helped preserve the trees even during droughts as well as from human destruction.

Surprisingly, with the same end in view modern environmentalists, in order to counter-balance deforestation and other evils of urbanization and industrialization, are paying much attention on creation of parks and conservation of wildlife sanctuaries. But these movements are not getting desirable strength and are not percolating to the common mass because the spiritual dimension, which Buddhism has, is totally absent. These movements can be made a public movement with the help of the spirit of Buddhism. It is evident from the Buddhist countries like Sri Lanka, Thailand, Bhutan, etc., and places like Laddhakh in India, that the respect for nature in the public has resulted in preservation of animals, birds, plant and trees. They are still living with nature in a harmonious way.

It seems relevant to mention here that the Buddhist theory of nature conservation is cosmo-centric in character. Here man is considered as the master of neither nature, nor nature is treated as a slave or a thing to be exploited for human pleasure or enjoyment. One is not tend to think for self-interest or the interest of human kind, rather it is out of respect and reverence for trees we protect or conserve it. Though this act is not purely selfless, the philosophy behind this, as said earlier, is to fetch spiritual merits for future or sublimated next birth. However, this remote self-interest, in the long run, went underground and the work of nature conservation was translated into the habits of Buddhist community. Thus, the so-called anthropocentric activity is converted into eco-centric or nature-centric as the impetus behind such actions now, is not the self-interest but a regard or the gratitude towards trees or plants. This theme is beautifully expressed in *Petavatthu* wherein it is said, "We should not even break the branch of a tree that has provided us shelter.[14] In Pali texts, trees are called 'vanaspati' which literally means 'lord of the forest'.[15] This is also a way to honour the trees and protect them from human atrocities. The huge trees of sal and fig, etc., are considered pious and are acknowledged *Bodhi* trees as the previous *Bodhi-sattvas* have meditated under their shadow.[16] The moral precepts that emanate from this Buddhist attitude are positive and cosmo-centric in nature. These love-kindness (*karuna*), joy *(mudita)*, friendliness (*metta*), eco-friendliness in the present context, as they have been formed after taking into consideration the well-being of nature.

Nature Management

Normally, man is dependant on nature for his basic needs such as food, clothing, shelter, medicine, etc., and nature supplies these without any trouble. But, when nature is exploited for nursing the unbridled greed and wasted mercilessly, eco-crisis emerges. In order to prevent this, a crisis management of natural resources is needed. This management of nature implies two types of action plan: (a) need-based use of natural resources, and (b) frugality.

Need-based Use of Natural Resources

Today, we are living in the age of consumerism, which has given rise to energy crisis and pollution problems. Due to vast hoarding and methodical destruction of large quantity of natural resources to feed our greed and ever evolving wants, the vast resources of fossil fuel and petroleum, which took millions of years to form, have come to the point of near exhaustion. Within forty years, Americans alone have consumed the same quantity of natural resources as had been consumed by all humanity in previous 4,000 years.[17] The dynamics behind this consumerism is human greed, lust, attachment and short sightedness, which prompt men to invent new commodities of comfort, varieties of artificial pleasure and exciting adventure. For this, nature is exploited. The jungles of wood are being cut and jungle of concrete is appearing in their place and the surface of the earth is changing rapidly.

Buddha is of the opinion that the environment (world) is the manifestation of intention (*citta*). Human intention (cravings), collectively, is the cause of existence and constant renewal of the universe.[18] Our intentions or thoughts are dependant on our mental make-up. If it is polluted with lust, hatred, and delusion (*lobha, dosa*, and *moha*), '*akasal citta*', it will translate itself into the external environment as complexes of physical life forms and material development based on exploitation of natural resources without moral restraint. In the internal environment it appears as complexes of unwholesome thoughts, greed, impulses, etc., which delude mind; we consider ourselves as independent individuals and in order to satisfy our desires and protect ourselves from the things we fear we exploit nature and manufacture such arms and weapons that are dangerous to the human life as well as to the ecosystem as a whole. Thus, all things and events of the universe are necessarily conditioned by the mental phenomena. This is the central tenet of the Buddhist Environmentalism. Due to ignorance and passion one fails to differentiate between need and greed. With a deluded mind one enjoys the pleasure of sense-object-contact. Delusion associated with greed results in environmental problems, as for satisfying one's greed natural resources are mercilessly exploited and man suffers. Buddha has repeatedly warned against it. In *Majjhuma Nikaya* the man deriving pleasure out of sense-object-contact has been compared to a leper who is scratching his wounds and heating it over fire and deriving momentary pleasure thereby, which ultimately worsens his wound and augments his trouble all the more.[19] In the same way a dog lustily gnaws at a fleshless peace of bone and gains ultimately nothing but pain and suffering.[20] *Dhammapada* categorically says that the fools not realizing the inherent painfulness of apparently pleasant object of the world, enjoy them like honey.[21] Similar is the case of present environmental problems. Mankind has exploited the nature for his pleasure and greed to such an extent that environment is polluted with disastrous consequences.

Buddha has repeatedly advised not to nurse greed, as greed is the cause of all evils and suffering.[22] With the end of greed all sorts of suffering is evaporated (*Tanhakhayo sabba dukkham jinati*).[23] He has asked to uproot the craving and crush the greed, and then only can our present problems be solved (*Tanha vippahanena sabbe chindantibandhanam*).[24] In order to eradicate the greed it is essential to develop the positive quality of contentment (*Santutthi*). For Buddha, contentment is the highest wealth.[25] A contented man is always happy with what he has got and his vision is non-exploitative. Hence, with limited wants he is neither harmful to nature nor creates pollution by his acts. Thus for Buddhism a person leading a simple life with a few wants is considered to be an exemplary character.[26]

Basic needs are decidedly much less than unbridled greed. For the satisfaction of the former, nature will not be exploited as in the case of the latter. If humankind, as a part of nature, works in harmony with it and uses its resources for the normal sustenance, damage to the system will be minimal. And this damage will be repaired automatically in its natural course. Hence, there will be no ecological or environmental problem.

Frugality

Another problem that is born out of ostentatious consumerism is pollution. With the development as well as for nursing our greed, human activities assumed such enormous dimensions that life supporting system could no longer sustain these. Accordingly, the waste generated through human activities was much more than the system could absorb or assimilate. And this has resulted in pollution. Buddha has deplored wastefulness.[27] For him those who waste the natural resources are derided as wood-apple eaters. In *Anguttar Nikaya* he says that a man in order to eat ripened apples shakes the branch of an apple tree. Many apples ripe as well as unripe fall. The man collects only the ripe apples and leaves the rest to rot.[28] The rotten apples in few days will start stinking and thus will create air pollution. Buddha has condemned such a wasteful attitude. For him it is an anti-social as a well as criminal activity. This problem of waste and in turn pollution can easily be solved through Buddhist *Madhyam Marga*, middle path. In the present situation, miserliness[29] and wastefulness[30] are two degenerate extremes. One should avoid these two extremes and adopt the middle path of frugality while using the resources of nature.

In Buddhism, frugality has been recommended as a virtue in its own right. *Pamsukulikanga* and *Tecivarikanga*, the first two of the 13 *dhutangas*[31] prescribed by Buddha, are the example of frugality in using clothes. *Pamsukulikanga*[32] recommends to utilize the clothes which are thrown out on the road side, after properly washing and mending, if it is useful. *Tecivarikanga*[33] suggests a monk not to possess more than three clothes at a time. These two *dhutangas* help a poor person lead a tension free life, and to an affluent curtails unlimited wants by generating in him contentment, and above all it helps recycle the waste material and lessens pollution in turn. Explaining the method of thrifty use of robes, Ananda says: "After receiving new robes the old robes are used as coverlets, the old coverlets as mattress covers, the old mattress covers as rugs, the old rugs as dusters, and the old dusters are used to repair the cracks of the floors and walls by kneading them with clay."[34] And thus, nothing is wasted. This example suggests that if we recycle the waste material and adopt frugality as a habit we will be able to prevent energy crisis and pollution, as well as save the natural resources from their misuse. For that every individual will have to adopt a simple moderate lifestyle based on contentment and frugality.

Protection of Nature

Here a question arises: from whom nature has to be protected? In fact, it is human beings who can maintain or mutilate the cosmos, as man has the capacity to think and invent such mechanism which can influence the course of nature. Nature has its own laws;[35] it has its own course of action. It is man who compels nature to violate its laws and change its course. In turn, nature retaliates adversely. It is simply a reciprocity.

Here it seems relevant to mention that man is one of the evolutes of nature. Since we are the highest (not the best) evolutes we have the ability to affect and influence the nature. At the same time this placement, as the 'highest evolutes', has also given us the responsibility to protect and promote the entire cosmic existence for two reasons: (a) if cosmic existence is not maintained we will also be destroyed, and (b) we both (cosmos and man) are equally important and both are interdependent. But due to greed, hatred and delusion our relation with nature and other beings has got strained. For our enjoyment and greed we adopt an aggressive attitude towards nature. Buddha has strongly opposed this attitude. He is of the opinion that for wanton destruction or even for need satisfaction, nature should not be spoiled or wasted as it upsets the vital balance of nature. This idea has been beautifully expressed in the conversation between Buddha and his disciple Maha Mogallan. When monks could not collect daily food as charity due to acute famine venerable, Maha Moggallan said to Buddha, "Lord, alms-food is hard to get in Vairanjna now. There is a famine and food tickets have been issued. It is not easy to survive even by strenuous gleaning. Lord, the undersurface of this earth is rich and as sweet as pure honey. It would be good if I turn the earth over. Then, the *bhikkhus* (monks) will be able to eat the humus that water plants live on."

"But Moggallan, what will become of the creatures that depend on the earth's surface?"

"Lord, I shall make one hand as broad as the Great Earth and get the creatures that depend on the earth's surface to go on to it. I shall turn the earth over with the other hand."

"Enough Moggalan, do not suggest turning the earth over. Creatures will be confounded."[36]

This is to my mind a warning to modern scientists who are engaged in inventing new methods to exploit nature in different artificial ways. Buddha is not in favour of disturbing the natural function of the world in any situation. He is against exploiting it even to feed the basic needs. In order to cultivate non-aggressive attitude Buddha has propounded a set of ethical code of conduct. They are as follows:

(i) *Panatipata Veramani* (to abstain from destruction of life);

(ii) *Adinnadana Veramani* (to abstain from stealing);

(iii) *Kamesumicchacara Veramai* (to abstain from sexual misconduct);

(iv) *Mishawaka Veramani* (to abstain from telling lies); and

(v) *Sura Majjapo Veramai* (to abstain from taking intoxicating thing). [37]

This is *Panca sila*, followed by Buddhist householders or laypersons. Buddhists and monks have to follow stricter precepts. The first precept is non-injury to life. Buddha has advised that one should abstain from such activities in which there are possibilities to harm any living creature even unintentionally. That is why, Buddha made a rule not to travel during rainy season as worms and insects that come to the surface in wet weather may be injured.[38] Once Buddha objected a monk, who was to set on fire his clay house in order to give it a fine finish, very strongly as in doing so many insects would be burnt in the process.[39] The same concern for non-violence prevents the monks and nuns from drinking unstrained water.[40] Though it is a sound hygienic habit, but reason behind it is sympathy and compassion *(karuna)* for all creatures of the universe. Buddha has prescribed the practices of friendliness *(metta)* and compassion *(karuna)* for all creatures timid and bold, long and short, big and small ... born and awaiting birth.[41] This is a positive aspect of Buddhist *ahimasa*, which presupposes cultivation of compassion and sympathy for all living beings.[42] Buddhist ethics springs out of love, sympathy, compassion and respect for all life. With the same end in view

Buddha has advised his lay followers (householders) not to earn their livelihood through wrong and immoral means. For him there are, broadly, five trades that should not be accepted and adopted for earning wealth or to maintain the expenses of one's family members. They are trading in weapons, in breathing things, meat, liquor and poison.[43] Buddha has beautifully summarized his moral precepts in a very simple statement of *Dhammapada:*

> *Not to commit any sin, but to practice all good,*
> *Keep the mind and heart pure, this is the teaching of (all) Buddas[44].*

Buddha has prescribed 13 *dhutanga* practices in order to curb the craving and greedy urges of the deluded mind.[45] Today, these may be prescribed to the affluent, statesman, policy makers, scientists and administrators, etc., as yoga therapy.[46] Observance of these will not only purge defiled mind and change their perception towards nature but also generate compassion, friendliness and compassion for fellow living beings as well as for nature and natural phenomena.

But no important change in ethics can be accomplished without any internal change in our mental and intellectual emphasis, loyalties emotion and conviction. This is the reason that the present environmental movements have failed to translate its basic principles into human conduct. For Buddhist *Paticcasamuppadavada*, ignorance is the root cause of present predicament. It is the first link of *Dvadasa Nidana*, which influences the life process, but is itself conditioned by the sensual desires, ill-will, laziness, agitation and fear of commitment *(panca nivaran)*, which are in turn conditioned by unskilful conduct *(akusal kamma)*. Interestingly, these twelve links are arranged clockwise. If we go forward it results in present crisis (suffering *or jara-marana)*. If it is followed in reverse order, it provides us solution of the present crisis. *Paticcasamuppadavada* says: "That being this, this comes to be....That being absent, this is not."[47] Hence, with the cessation of ignorance the consequent cessation of each following links *(nidana)*[48] and the present environment problems will also cease to be. Hence, the prime purpose of the Buddhist path is to regain the lost vision *(sammaditthi)*, regarding sameness and interdependence of man and nature, his inherent goodness, etc., through destruction of ignorance. This deep-seated ignorance can only be destroyed by meditative insight *(panna)*, and thus by the rise of wisdom, craving and ignorance both are destroyed.

Finally, entire Buddhist environmentalism can be expressed in the following two verses:

Antojata bahi jata jataye jatita paja
Tam tam Gotam pacchami ko ime vijataye jatam" ti.[49]

It means that entire human kind *(paja)* is bound by matted knots internally (by greed, delusion, aggressiveness, etc., i.e., mental pollution and externally (environmental pollution). "It is called inner tangle and outer tangle because it arises (as craving) for ones own requisites and another... and for the internal and external bases (for consciousness)."[50] This tangle goes on arising again and again, up and down like the branches in the bamboo thickets. In the same way, entire living beings are intertwined and interlaced by the mental and physical pollution. Gautama, the Blessed One, who is capable of disentangling this tangle, says:

Sile patittayo naro sapanno. Cittam pannancabhavayam
Atapi nipako bhikkhu, so ime viyataye jatm" ti.[51]

It means, a wise man that is well established in virtue develops consciousness and understanding. Then, *O Bhikkhu*, the person who is ardent and sagacious will succeed in disentangling this tangle.[52]

Notes

1. Avinash Kumar Srivastava "What is Living and What is Dead in Buddhist Philosophy," in *Nalanda and Buddhism*, Research, Vol. VII, Nava Nalanada Mahavihar Nalanda, 2002, p. 242.
2. Ajit Kumar Sinha, "A World View, Through a Reunion of Philosophy and Science," published by The Library of Philosophy, Calcutta (1959), p. 1.
3. *Majjhama Nikaya*, I, 167, pp. 247-249.
4. *Digha Nikaya (mahaparinibbana sutta)*, II, pp. 138-40.
5. *Majjhima Nikaya*, I, 504.
6. Ibid., 17.
7. *Vinaya Pitaka*, I, 8.
8. Ibid., 352.
9. *Samayutta Nikaya*, IV, 373; *Majjhima Nikaya*, I, 118.
10. Visuddhimaggo of Buddha Ghosh, *Dhutanga Niddesa*, VIII Dhutanga; Path of Purification, translator, Bhikku Nanmoli, Buddhist Meditation Centre, Singapore (1997) pp. 72-74.
11. Ibid., pp. 75-76, IX the Dhutanga.

12. A.K. Srivastava, "Dhutanga Niddesa, Middle Path and Modern Man", *Published* in *Buddhism in Universal Perspective,* Nava Nalanda Mahavihar, Nalanda, 201 pp. 144-152.

13. *Smayutta Nikaya,* I, 33; *Jataka,* I, 199.

14. *Petavathu,* II, 3, 9.

15. *Samayutta Nikaya,* IV, 302.

16. *Digh Nikaya* I-Atta kattha 1.

17. *The Waste Makers,* Orient Longman, London (1961), p. 195.

18. "Cittena niyati loke, cittena parikassati.Cittassa eka dhammassa, sabbeva vasa manvabhuti", *Samayutta Nikaya,* I, 39.

19. *Majjhima Nikaya,* I, 507.

20. Ibid., 364.

21. "Madhuva mannati balo yava papam na paccati Yada ca paccati papam atha balo dukkam nigacchati", *Dhammapada,* p. 69.

22. Dr. Kala Acharya, *Buddhanusmriti,* "A Glossary of Buddhist Terms", Somaiya Publication Pvt. Ltd., Mumbai (2002), pp. 177-178.

23. *Dhammapada,* 354.

24. *Majjhima Nikaya,* II, 98, *Dhammapada.*

25. *Santutthi paramam dhanam, Dhammapada,* 204.

26. *Anguttara Nikaya,* IV, p. 220, 229.

27. *Dhammapada Atthakatha* III, 129.

28. *Anguttara Nikaya,* IV, 283.

29. *Dhammapada Atthakatha,* I, 20.

30. Ibid., III, 129.

31. *Path of Purification,* pp. 59, 62-66, *op. cit.*

32. Ibid., pp. 63-66.

33. Ibid., pp. 65-66.

34. *Vinaya Pitaka,* II, 291.

35. According to Atthasalini there are five natural laws (Pancaniyamadhamma) They are *utuniyama* (season Law) *bijaniyama* (seed law), *Cittaniyama* (mind law) *Kammaniyama* (action law) and *dhammaniyama* (phenomenal universal law) Atthasalini, p. 854.

36. *Vinaya Pitaka, Suttavibhanga,* Para I.

37. *Majjhima Nikaya,* III, 45-50.

38. *Vinaya Patika,* I, 137.

39. Ibid., III, 42.

40. Ibid., IV, 125.

41. *Digha va ye mahanta va.... bhuta va sambhavesiva... nannetha katthaci nami karrci,* Suttanipata, (Mettasutta), pp. 145-146.

42. *Digha Nikaya*, I, 4.

43. *Anguttara Nikaya*, V, 177.

44. *Sabha papassa akaranam kuslassa upasampda Sacittapariyodapanam etam Buddhamas sasanam, Dhammapada*, 183.

45. *The Path of Purification*, p. 59, *op. cit.*

46. Avinash Kumar Srivastava, "Dhutanganiddesa Middle path and Modern man", in *Buddhism in Universal Perspective*, Nava Nalanada Mahavihara, Nalanda (2001), p. 151.

47. *Smayutta Nikaya*, II, 28.

48. *Vinaya Pitak*, I, 1.

49. *Visuddhimaggo*, p. 1, *Samuyatta Nikaya*, I, 13.

50. *The Path of Purification*, p, 1, *op. cit.*

51. *Visuddihimaggo*, p. 1, *Samuyatta Nikaya*, I, 13.

52. *The Path of Purification*, p. 2, *op. cit.*

III

Environment and Literature

17

Nature and the Poet: Observations on Tulasidasa and Valmiki

Prabhat Kumar Pandey

Then God said, "Let us make man in our image, in our likeness, and let them rule over the fish of the sea and the birds of the air, over the livestock, over all the earth, and over all the creatures that move along the ground."

Genesis 1:26

Pra vatah vanti patayanti vidyu atah ut oshadhih jihate pinvate
Sva 1 riti svah ira vishvasmai bhuvanaya jayate yat parjanyah
prithivim retasa avati.

Rigveda, Parjanya Sukta 4

(To bring rain the winds blow, lightings flash, vegetations sprout, raindrops fall from the sky, and the earth is enabled to do good to the world when the clouds irrigate the earth.)

Ashvathah sarvavriksanam devarshinam ca Naradah I
Gandharvanam chitrarathah siddhanam Kapilo muni II

Bhagavadgita X: 26

(Among the trees I am *ashvath*, Narad among divine *rishis*, Chitrarth among gandharvas and Kapil Muni among *siddhas*.)

For the Indians, all nature is divine... Man is no longer at the centre of life. He is no longer that flower of the whole world... He is mingled with all things, he is on the same plane with all things, he is a particle of the infinite, neither more nor less important than the other particles of the infinite... because Nature was (with them) something other, something more than a mere backdrop. Because man, though divine, was no more divine than that from which he sprang (Elie Faure, *History of Art II*).

Tulasidasa had modestly said that he had composed the *Ramacharita Manas* for *svantah sukhaya* (for his own contentment) but the work has great social importance, for every couplet is a pearl of wisdom. His treatment of 'nature' is unique: realistic description with a moral lesson. Valmiki's treatment of nature is beautiful, sensuous but without moral note. Tulasi wrote the *Manas* on pattern of the *Ramayana*, following its chapter division, and at times giving descriptions of the same natural phenomena, especially the *ritus* (seasons). This paper shows the treatment of nature by these two great poets and how they differ from each other. Since both of them describe seasons, so the focus is on it, though some other nature descriptions are also included. I will take up their treatment of *varsha* and *sharad*. Unlike the usual three seasons that we use today ancient Indian system divided the year into six *ritus: vasanta* (spring), *greeshma* (hot season), *varsha* (the rainy season), *sharad* (autumn/fall), *hemanta* (winter), and *shishir* (the cool season). They are personified, addressed in *mantras* and worshipped by libations in the Vedas.

Let us see how *varsha ritu* is described in the *Kishkindha Kanda*. Sita is abducted and Rama is depressed. First, he describes the season and how peacocks are dancing gayly:

Lachhiman dekh mor mana nachat barid pekhi I
Griha birati rata harash jas bishnu bhagat kahun dekhi II

(Lo Lakhan, how dance peacocks seeing clouds,
Like the Vishnu devotee transporting the ascetic.)

But, in the next verse, human aspect of the god is shown even though he is *Maryada Purushottam Rama*. In a reverse role, he is scared by the thunder, rather unmanly:

Ghana ghamanda nabha garajat ghora, priya hina darapat mana mora I
Damini damak rah na ghana mahin, khala kai preet jatha thir nahin II

(Thundering clouds scare me, sans Sita,
Lightening lasts not, like love of the wicked.)

In the second line transitory lightening is aptly compared with the fickle love of the wicked. Then, he describes the low hanging clouds bringing rain:

Barasahin jalad bhumi niaraiyen, jatha navahin budh bigha payen I
Bunda aghat sahahin giri kaisen, khala ke vachana sant saha jaisen II

(Clouds bow to rain, as the learned bow with modesty,
How hills bear rain lash, as saints' words of wicked.)

Analogies in both the lines are striking but the second line is more so, bringing to our mind the English metaphysical poets. The second analogy is a conceit: how can rain drops hurt stone?

The saint is stone and the wicked's words hurt him not. Stone is impervious, so are the saints. The rivers overflow in rainy season and so some men. But how are rivers like men:

Chhudra nadi bhari chalin torai, jas thorehun dhana khala itarai I
Bhumi parat bha dhabhar pani, janu jiva hi maya lapatani II

(Small rivers swell, like self-conceited with little money,
Water soils touching earth as *Maya* smearing life pure.)

Raindrops are pure until they fall onto earth and so happens with the *jiva*. The moment baby comes into the world the Vaishnavi *vayu* permeats him with *maya*. The ponds fill slowly:

Samihin samiti jala bharahin talava, jimi sadguna sajjana pahin ava I
Sarita jala jalanidhi mahun jai, hoi achala jimi jiva hari pai II

(Slowly water fills the pond, as virtue makes the noble,
Rivers stay put in the sea, as men would realizing Hari.)

Hari *bhakti* gives stasis of the deep like *vairagya*, freedom from passion and attachment. Many small drops fill the pond as noble virtues make a man perfect. Tulasidasa wrote during the time of social turmoil, spiritual decay, and Muslim repression. In the *Uttar Kanda* of *Kavitavali* he has described the rule of Kali (evil) in Kashi under the heading "Epidemic in Kashi". Several sham sects and *mathas* had cropped up eclipsing genuine faith. How well he says:

Harita bhumi trina sankula samujhi parahin nahi panth I
Jimi pakhanda bada ten gupta hohin sadgranth II

(Green grass all around has concealed paths,
Like sham sects eclipse good scriptures.)

In the next verse Tulasi humorously compares Veda reciting with croaking frogs but in the second line the analogy is correct:

Dadur dhuni chahu disha suhai, Beda padhahin janu batu samudai I
Nav pallav bhaye bitapa aneka, sadhak mana jasa milen bibeka II

(Croaking frogs sound good like Veda reciting pupil's,
New leaves looking good, like *sadhak* getting wisdom.)

The *adi kavi* (first poet) Valmiki in *Kishkindha Kanda* has thus described the croaking frogs:

*Svanairdhananam plavagah prabudhha vihaya nidram
chirasannirudhham* I
Anekarupakritivarnanada navambudharabhihata nadanti II

38:38

(Hearing thunder many hibernating creatures come to life, Frogs of many colours, forms and sounds are croaking happily bathed in rain.)

The description is apt but has no lesson in it. Valmiki humanises nature in the following verse but Tulasi relates her to life:

Navamasadhritam garbham bhaskarasya gahabhastibhih I
Pitva rasam samudranam dyauh prasutam rasayanam II

28:3

(The maiden drinking seawater has conceived for nine months. But she has conceived water.)

Tulasi also has described autumn and relates it with life:

Arka javas pata bina mayau, jas suraj khala udyama gayau I
*Khojat katahun milai nahin ghura, karai krodha jimi dharamahin
duri* II

14:2

(*Madar* and *javas* denuded as kingdom sans wicked,
Not a speck of dust at all, like anger banishing *dharma*.)

Rain brings life. Earth is green with rich crop. The bounty of nature is seen everywhere. But in the second line assembly of the arrogant as glow-worms is arresting:

Sasi sampanna soi mahi kaisi, upakari kai sampati jaisi I
Nisi tama ghana khadyot biraja, janu damhinha kar mila samaja II

14:3

(Crop-rich earth so gracious like benefactor's wealth,
Glow-worms in murky night like gatherings of arrogants.)

Descriptions of *varsha ritu* continues but every phenomenon of it has certain moral. First, the rivers overflowing their banks, then wind pushing clouds away:

Mahabrishti chali phuti kiarin, jimi sutantra bhayen bigarahin nari I
Krishi niravahin chatur kisana, jimi budh tajahin moh mada mana II
4

Kabahun prabal bah marut janh tanh megh milai I
Jimi kaput ke upajen kula saddharma nasahin II
15(a)

(Rain overflows bed, like women falling unrestrained,
Peasant weed out field, like the wise weeding arrogance.
Fast winds obliterate clouds
Like bad son blemishes clan's name.)

It is the son who either maintains and increases family honour or by his evil deeds spoils it through a commonplace thing *kusal*, a common holy wild shrub. Blooming of *kasa* (corruption of *kusha*) announces end of rainy season:

Barasa bigat sarad ritu aai, Lachhiman dekhahu param suhai I
Phulen kasa sakal mahi chhai, janu barasan krit pragat budhai II

(Lo Lakhan, rain over, *sharad* has come,
Kasa has covered earth as aging rain.)*

We may inform that *kasa* flower (not exactly flower) is white hence comparison with aging. Valmiki also describes *kasa* in his description of *varsha*:

Navairnadinam kusumaprasairvyadhupamanairmridumarutena I
Dhautamalaksaumapataprakashaih kulani kashairupashobhitani II
30:51

(Riverbanks with breeze look beautiful with flower laugh and clean washed silky fresh *kasa*).

Here beauty of *kasa* is well compared with spotless white silk but it does not suggest aging of *varsha*.

During *sharad* because of sun's heat, water dries and both Valmiki and Tulasidas describe this phenomenon but moral is found only in the latter. Valmiki has personified the river with a blushing maiden which is a striking simile but it has only literary beauty:

---

* English rendering is my own.

Darshayanti sharannadyah pulinami shanaih shanaih I
Navasangamasavrina jaghananiva yoshitah II

30:58

(In *sharad* gradually receding water exposes riverbanks, as in the first sexual congress the maidens are compelled to gradually expose their privates.)

Tulasi has described this phenomenon thus:

Udati agast panth jala sosha, jimi lobhahin soshai santosha II
Sarita sara nirmal jala soha, santa hridaya jasa gata mada moha II
Rasa rasa lukh sarita sara pani, mamta tyaga karahin jimi gyani II
Jani sharad ritu khanjan aae, pai samay jimi sukrit suhaae II

(Star Agastya soaks water, as contentment greed,
Clear pond water, like saints sans passion pride.
Slowly drying water, the wise renounce passion,
Khanjan comes in *sharad*, as good deeds in time.)

Tulasi does not eroticise nature. He relates every phenomenon of nature, in this case season, with life. Thus drying water is like contentment, soaking greed and clear water in the pond is like the heart of saints without *raga* or *dvesha*, which is *vairagya*. With coming of *sharad* many activities begin which were stopped because of rains, be it invasion, commerce, penace or begging:

Chale harashio taji nagar nripa tapas banik bhikhari I
Jimi hari bhagat paai shram tajahin ashrami chari II

(King, ascetic, trader, beggar leave the town,
As Hari devotee renouncing four ashrams for all.)

Sharad is also the flowering season with many flowers blooming, black bees humming and birds cooing. This is the season of rejoicing, for nature is rejoicing but even in such time some may grieve. Who can he be:

Gunjat madhukar mukhar anupa, sunder khaga rav nana rupa I
Chakrabaka mana dukha nishi pekhi, jimi durjan par sampat dekhi II

(Black bees hum, many birds gayly coo,
Chakava grieved in dark, as wicked seeing other's wealth.)

Chakava is unhappy because of absence of moon but the wicked grieve at seeing prosperity of others. When the *sadhak* realizes Hari, he needs nothing more and is contented as fish in the deep:

Sukhi meen je neer agadha, jimi hari saran na ekahu badha I
Phule kamal soh kar kaisa, nirguna brahma sagun bhaen jaisa II

(Fish in the deep happy, like one in Hari's refuge,
Blooming lotus in the pond as *nirguna* Brahma with attributes.)
The second line is striking. Tulasi suggests that blooming lotus in
the pond are so charming as though *nirguna* (attributeless) Brahma has
become *saguna* and *nirguna* Brahma *upasana*. Craving of *chatak* bird is a
stock image in Sanskrit and Hindi poetry but here its thirst is given a new
dimension; it is compared with an atheist:

Chatak rahat trisha ati ohi, jimi sukh lahai na sankar drohi I
Saradatapa nisi sasi apaharai, sant daras jimi patak harai II

(Chatak thirsty craving as Shankar's for unhappy,
Sharad moon banishing heat, as *darshan* of a saint.)
Sharad is a pleasant cool season. It was in this season that Lord
Krishna performed *maharasa* (the great dance with thousands of *gopis*).
To us mortals, sharad replaces heat with cool, soothing like *darshan* of a
saint. Chakore is in love with the moon and constantly grazes at it. The
two are inalienable and are referred to as eternal lovers. It is a stock image
for perfect couple. But what can human learn from it? Let him fix his
gaze at the Lord:

Dekhi Indra chakore samudai, chitavahin jimi harijan hari pai I
Masak dansa beete hima trasa, jimi dvija droh kiyen kula nasa II

(Chakore gazing at moon, like Hari devotees at the Lord,
Mosquitoes dead by cold, family perished by Brahmin's wrath.)
The virtuous *guru* removes all our doubts and confusion as during
the sharad many rainy creatures are destroyed:

Bhoomi jiva sankul rahe gaye sharad ritu pai I
Sadguru milen jahin jimi sansay bhram samudai II

(Rain-born creatures destroyed by Sharad,
As guru doth remove doubt illusion all.)
After the rain sky becomes clear without clouds whose beauty is
described thus by Valmiki:

Vyaktam nabhah shastravidyautavarnam krish pravahani narijalani I
Kahlarashitah pavanah pravanti tamo vimuktashca dishah prakashah II

(The sky is clear bright like the sword on a whetstone, rivers flow gently; the sun removes breeze carry lotus fragrance, the sun removes darkness from all quarters.)

Sharp edge of the sword is an apt analogy but see how Tulasi relates the same phenomenon with a moral:

Bin dhanu nirmal msoh akasa, harijan iva parihari sab aasa I
Kahun kahun brishti saradi thori, koda eka pav bhagati jimi mori II

(Cloudless sky elegant as the realized Hari devotee,
Rain in *sharad* rare, as rarely one reaches me.)

In *Aranya Kand* of the *Ramacharitamanas* also, Tulasi has given some description of nature with a moral. Lord Rama has reached the Pampa lake whose beauty is described thus:

Santa hridaya jas nirmal pari, bandhe ghat manohar chari I
Janh tanh piyahin bibidh mriga nira, janu udar griha jachak meera II

(Pampa water clean as hearts of saints, lovely ghat four,
Several creatures drink, as beggers at the donor's door.)

The world consists of Jiva, Maya and Brahma. Jiva and Brahma are one separated by maya as seen above in the verse 13:III.

Bhumi parat ma dhabar pani, janu jivahin maya lapatani II

Maya blinds man so he cannot find Brahma and this is expressed thus by Tulasi:

Purain saghan ota jala begi na paia marma I
Mayachhanna na dekhiai jaise nirguna Brahma II

(Lotus leaves conceal pond water,
Like Maya hiding nirguna Brahma.)

We have seen above how lotus flowers are compared with Nirguna Brahma becoming Saguna and here lotus leaves conceal like Maya. The interrelationship of Jiva, Maya and Brahma is described thus by Tulasi when Rama is banished to the forest:

Aage Rama Lakhanu bana pachhe, tapas besh birajat kachhen I
Ubhaya beech Siya sohanti kaise, Brahma jiva bich maya jaise II

(Rama in front and Lakhan at back, pretty as ascetic,
Sita in between like Maya between Brahma and Jiva.)

Similar verses are also found in Tulasi's *Geetavali*, song 50 describing Chitrakut during the monsoon:

Chahun dishi bana sampanna bihanga-mriga bolat sobha pavat I
Janu sunaresa desa-pur pramudita praja sakal sukh chhavat II

(All around forest rich, birds and deer grace,
As the thriving subjects under a noble king.)

Mandakinihi milat jharana jhari jhari bhari bhari jala aachhe I
Tulasi sakal sukrit-sukh lage mano Rama-bhagati ke pichhe II

(Clean cascade water trickles in the Mandakini,
Tulasi, good deeds all so follow *Rama bhakti*.)

The aphorisms that wisdom and wealth make one modest, fruit-laden trees bow with weight. We have above the modest scholar in verse 13,II. Her is an example of the noble rich who is always humble:

Phala maran nami bitap sab rahe bhumi nijrai I
Para upakari purush jimi navahin susmpati pai II

Aranya Kand, 38:40

(The fruit-laden tree coming near earth,
As the benevolent bow getting right wealth.)

Though Rama loved Janaki deeply yet in *Aranya Kand* when explaining Narad why he as Vishnu had stopped him from marrying, he speaks about the evil influence of women. In the following verses he compares women with season, night and animals:

Kama krodh mada matsar meka, inhahin harashaprada barasa eka I
Durbasana kumud samudai, tinha kahn sarad sada sukhadai II

(Lust wrath envy pride, frogs pleased by rain-women,
Evil passions hunch of lotus ever pleased by sharad.)

Here the women are compared with the rainy season, which delights the frogs that are different flames of mind or sins. In the following verse the women is compared with darkness, so does sin and with the fishing rod of woman one kills wisdom, piety, truth and strength. She destroys all that make one virtuous:

Papa uluka nikar sukhakari, nari nibida rajani andhiari I
Budhi bala sila satya saba mina, banasi sama triya kahahin prabina II

43:4

(Women-murk delight a pack of sin owls,
Woman-fishing-rod kills wisdom, piety, truth, and strength.)

To conclude, Valmiki is considered as the first poet and his *Ramayana* the original book of the story of Rama on which several versions in different Indian languages have been composed.** Tulasi himself acknowledges his indebtedness to him but in one respect he supersedes him, in writing didactic poetry. Valmiki's description of nature is superb. He humanizes nature well with personifications but he does not relate her to life. Tulasi has devoted less space to description of *varsha* and *sharad* than Valmiki but his descriptions always carry a moral.

** Valmiki uses Nature as *alambana* (independent description) but Tulasi uses her as *uddyipana* (to express emotion/moral).

18

Nature in Hindi *Chhayawadi* Poetry: An Instance of Jai Shankar Prasad's *Kamayani*

Sanjay Kumar

The span of 15 years from 1920 to 1935 saw the rise and growth of a new kind of poetry in Hindi known as *chhayawadi* poetry. The chief exponents of this poetry are Jai Shankar Prasad, Suryakant Tripathi 'Nirala', Sumitra Nandan Pant and Mahadevi Verma. It was Jai Shankar Prasad who heralded this movement with the publications of *Aansu* in 1920, inventing and fashioning a new poetic idiom and diction to express new modes of experience and new relations of the individual consciousness to itself. This movement culminated later in the poetry of Pant and Mahadevi Verma. What distinguishes their poetry from the poetry of earlier era is, among other things, their special relationship with nature. Nature becomes a prominent subject in their poetry, so much so that they have often been called nature poets. They explore the possibilities of a union, by means of imagination, between mind and nature, in a recirocity that redeems the world of ordinary experience. They discover reciprocity between nature and their own minds in which the two agents are equal in initiative and power. They portray nature comprehensively in all its aspects, the beauty and splendour of nature; nature as a medium for expressing poet's emotions and feelings; nature as a stimulus for poet's emotions and feelings; nature as being alive and rational capable of sentience, often acting and behaving as human beings and thus interacting with the human world; nature as a repository of moral and spiritual truths; and nature as a manifestation of the Universal Divine Spirit.

Rooted in the Indian philosophical tradition of *advaita* and mysticism, the *chhayawadi* poet believes in an immanent and transcendent Divine Spirit which has not only created the universe and all forms of life but also pervades them. This all-pervasive presence of the Divine in all entities makes the *chhayawadi* poet believe in the organic unity of all creation *Sarvam Kahlu Idam Brahma* (all is Brahman). All existence—earth, heaven, planets, gods, living and non-living objects—is conceived here as the parts of One Great Person (*Purusa*) who pervades the world, but also remains beyond it. Man and nature are parts of the Universal Divine Spirit and they remain inseparable (*abheda*) from it. Nature is perceived as a manifestation of the same Divine Spirit, which manifests itself as man. Thus, the *chhayawadi* poet sees an essential unity between man and nature. He does not treat nature as dead and insentient inorganic matter, but treats it as alive and rational and also sees it as inter-acting with the human world.

Jai Shankar Prasad too believes in the immanence of the Divine Spirit that pervades the entire creation and he calls it *Shiva* or *Great Shiva (Param Shiva)*. The Divine Shiva while being present in all beings—living as well as non-living—remains one and the same, unaffected and unchanged by the temporal and accidental particularity of these beings. It also pervades all the different beings in the same uniform way.

Sab mein ghul kar rasmay rahta vah bhav charam hai.

Kamayani, Anand, 18

Shiva is not only immanent but also transcendent and this immanent and transcendent character of Shiva is shown in his dance of destruction in the 'Darshan' canto of *Kamayani:*

Antarninad dhwani se purit, thi shunya-bhedini satta chitra;
Natraj swayam the nrityavirat, tha antariksh prahsit mukharit;
swar laya hokar de rahe taal, the lupta ho rahe dishakal.

Darshan, 107

Shiva is the creator and the material cause of nature/universe. Nature is only a creation and manifestation of Shiva. It comes into being when he wills and causes it, it emerges from Shiva, finally merges, and dissolves into him, it is a part of Shiva and remains inseparable from him.

Lila ka spandit ahlad, vah prabhapunja chitimay prasad;
Anandpurna Tandav Sundar, Jharte the ujjawal shram sikar;

Bante tara, himkar dinkar, uda rahe dhulikana se bhudhar;
Samhar srijan se yugal pad – gatisheel anahat hua naad!

Darshan, 108

Man also being a creation and manifestation of Shiva is not different and separate from nature. It is illusion (*Maya*) and ignorance (*Ajnana*), which delude him into believing in the separate and individual existence of all entities including his own self. But when he turns inward to the contemplation of the abstract and subtle, from particular and concrete manifestations, he becomes aware of the similarities between self (*Jiva*) and nature (*Prakriti*). This is the stage of *Sadavidya* when he experiences *Aham Idam* (I am this) where *Aham* is the self, conscious and rational, and *Idam* is nature. At this stage, duality between man and nature exists, but they are of equal importance.

The next stage is the stage of *Ishwartatva* where man experiences *Idam Aham* (This is I). The duality of man and nature ceases and man experiences oneness/identification with nature.

Sab bhedbhav bhulvakar, dukh-sukh ko drishya banata
Manav kah re ! yah main hun yah vishwa needa ban jata.

Anand, 119

Then passing through the stages of *Sadashivtarva* and *Shaktitarva* man finally reaches the stage of *Paramshivtatva* where he immerses and dissolves his self into *Param Shiva* and experiences oneness with him. *Kamayani* ends with Manu experiencing oneness with nature and with *Param Shiva*.

Yah chandra kirit rajat nag spandit sa purusha puratan,
dekhta mansi gauri lahron ka komal nartan;
pratiphalit hui sab aankhein us prem jyoti vimla se,
sab pehchane se lagte apni hi ek kala se;
samras the jara ya chetan sundar sakar bana tha,
chetanta ek vilasti anand akhand ghana tha !

Anand, 120

Let us now consider how these philosophical beliefs mould and shape the *chhayawadi* poet's perception of nature and how this perception gets manifested in the epic *Kamayani*. First, a word about *Kamayani*. Published in 1935, *Kamayani* is the last poetic creation of Jai Shankar Prasad. Written in the epic form *Kamayani* is an allegory in which the three characters—Manu and his consorts, Shraddha and Ida—represent Man (consciousness), Heart (love, compassion, kindness and faith) and Intellect respectively. It rejects the materialistic philosophy of life, which treats this universe/nature as nothing but an object of consumption existing solely at and for the pleasure of man. It represents journey from ignorance to enlightenment and bliss and this enlightenment comes with the awareness of the unity of all creation, the divine character of all beings and feeling of oneness with and among them. Of the two— Shraddha (heart) and Ida (intellect)—Prasad gives primacy to the former, for it is Shraddha which leads us on the path of deliverance and salvation. Intellect/rationality promotes ignorance, it creates endless desire and passion in man and makes him conflict ridden and strife-torn, it cannot give him peace and joy. We see in the case of Manu who under the influence of Ida becomes materialistic and seeks gratification of his desires and passions, first by raising an empire and then lording over it cruelly and wantonly exploring it. This mindless materialism and consumerism, however, does not satisfy him and leaves him more unhappy and dissatisfied than ever. This wanton exploitation ultimately leads to a revolt against him in which he is overthrown, wounded and left for dead. It is Shraddha who nurses him back to health, shows him the folly of the consumerist culture which has made him extremely self-centred and indifferent to the welfare and well-being of others, and leads him to enlightenment by making him experience the unity and oneness of all things and thereby, the unity with the Divine Spirit.

Born with an aesthetic sensibility, the *chhayawadi* poet always experiences pleasure and joy in the contemplation of beauty. He is irresistibly drawn towards nature as he sees in it beauty in perfection, abundance and infinite multiplicity, not to be found anywhere else. Nature for him is a perennial source of joy. He is deeply moved and stirred by the beauty of nature and it arouses a sense of wonder and astonishment in him. Enchanted by the beauty of nature, Prasad exclaims in his essay "Prakriti Saundarya":

O Nature Goddess! Salutation to Thee, your form is indescribable. The island, continent, peninsula, sea, river, mountain, city or the entire water-area is in your belly. In them is visible your charming beauty along with divine craftsmanship. What a marvelous creation at the bottom of the fathomless ocean, what wonder! How excellent! Swaying of coral creepers to the tune of waves, movement of oyster shells and other tiny creatures and multitude of wonderful creatures near unique creepers and vegetation and their wandering amid moving bushes with waves, such priceless gems at the bottom of unfathomable waters! And such charming scene![1]

He becomes a keen observer of nature and is drawn to even its smallest particles. He not only passionately and lovingly represents various objects and aspects of nature—flowers, plants, trees, mountains, rivers, falls, etc.—but also creates an integrated and unified image of nature. Such descriptions of nature are known as *alamban* (independent description) in Indian poetics. Such descriptions do not merely function as a backdrop to the human situation, but they stand on their own, becoming an end in themselves. Nature descriptions as *alamaban*, though to be found in all the *chhayawadi* poets, are not very common and frequent in their poetry. Pant's poetry is replete with such descriptions while in Prasad's poetry, such instances are few and they mostly occur in his early poetry. In the initial phase, in poems like "Urvashi" and "Vabhruwahan" nature arouses in him a sense of wonder and astonishment. It the second phase, in poems like "Pratham Prabhat", "Chandra", "Rajni" and "Indradhanush" this sense of astonishment gives way to a simple and passionate love of nature. In the mature phase, in an epic poem like *Kamayani* independent descriptions of nature are much less frequent, for nature is now well integrated into his vision of life and creation. Nonetheless, such descriptions are there and I will cite here a couple of examples. Prasad is a Shaivaite and appropriately, *Kamayani* begins with the description of the Himalayas, the sacred mountain and ends with the description of the blissful environs of Mount Kailash, the abode of Lord Shiva. Prasad paints a beautiful picture of the Himalayas spreading peace, tranquillity and joy all around, its lush green creeper and tree-covered body, its sweet falls, etc.

> *Anchal Himalaya ka Shobhantam lata kalit shuchi sanu sharir,*
> *Nidra mein sukh swapna dekhta jaise pulkit hua adhir!*
> *umad rahi jiske charno mein niravata ki vimal vibhuti,*
> *sheetal jhrano ki dharayen bikharati jeewan anubhuti!*
>
> *Asha*, 21

To cite another example of nature-description as *alamban*, Prasad evokes a beautiful image of dawn at the time of introducing Ida:

> *Prachi mein phaila madhur raga*
> *Jiske mandal mein ek kamal khil utha sunahara bhar parag*

Jiske parimal mein vyakul ho shyamal kalarav sab uthe jaag
Aalok rashmi se bune usha anchal mein aandolan amand
Karta, Prabhat ka madhur pavan sab or vitarne ko marand.

Ida, 70

Prasad describes the rising sun spreading its golden rays as a blossoming lotus full of golden pollen and sweet fragrance. With the sunrise, birds sitting on the green branches of trees start chirping as if they were singing paeans to their king. The first rays of light are perceived as the border of the *saree* of the young maiden, dawn. The breeze becomes restless to spread the fragrance of lotus-like sun in all directions. The poet here draws on visual, gustatory and olfactory senses to create the image of dawn.

We find similar independent descriptions of the sky, stars, clouds, dusk day, evening, night, etc., in *Kamayani*. The images of nature though mostly benign, kind, tender and beautiful are not always so. We also come across hard, harsh and terrifying images of nature as in the image of the great flood (*pralaya*) and the devastation caused by it.

Hahakar hua krandanamay kathin kulish hote the choor;
hue diganta badhir, bhishan rav baar baar hota tha krura
digdahon se ghoom uthe, ya jaldhar uthe kshitij tata ke!
saghan gagan mein bhi prakampan, jhanjha ke chalte jhatke
udhar garajati sindhu lahariyan kutil kal ke jalon si
chali aa rahi phen ugalti phan phailaye vyalon si!

Chinta, 14

The *chhayawadi* poet humanizes nature in his poetry; he endows nature with human attributes, capabilities, sensations and emotions. The images of nature humanized are most common *chhayawadi* poetry. Since the *chhayawadi* poet perceives nature as a living entity pulsating with life, ascribing human attributes to it poses no problem for him. He feels a kinship with nature and he often perceives nature behaving like a fellow human being. He portrays nature in different human roles a reveller playing and enjoying, a host welcoming guests, a beautiful woman, etc. Nature is also described in its various moods—happy sometimes, sad at other times, deprived and yet distressed at other times. This perception of nature as a sentient and rational being is the negation and denial of the anthropocentric view of the universe. In according nature (and by extension all living and non-living beings) a status equal to that of man, the *chhayawadi* poet denies the centrality of man and instead emphasizes the fact of mutual interdependence of all organisms and the need for their mutual coexistence. Such a cosmo-centric vision shows everything in this universe including man to be existing in its own right. Man is not seen as the lord and master of the universe, but just as one among many inhabitants not having any better or superior claim to exist and prosper at the cost of the others. Such a vision stipulates that man must not exploit and encroach upon the rights of others and that he should learn to inculcate a fellow feeling towards other living and non-living beings and to coexist with them. He should see the same Divine Spirit reflected in other entities which he sees in himself and such a vision would make him feel one and united with nature and other living beings.

In *Kamayani*, Prasad humanizes nature and its objects and describes their activities, transactions and emotions and feelings. All aspects and forms of nature—different seasons, dawn, dusk, morning, evening, day, night, plants, trees and creepers, flowers, birds and animals, rivers and seas, mountains, sky, etc.—are humanized and they are shown engaged in human activities. Nature is shown to pass through different stages of man—an innocent, guileless and fidgety infant to a lovesick youth engaged in courting and lovemaking to a suffering old widow. The objects of nature are portrayed variously as a drunk, a singer, a dancer, a painter, a hostess serving wine, a meditating ascetic, etc. Prasad has created in *Kamayani* a conscious, live and rational world of nature, which runs alongside the human world and often interacts with it.

Prasad often depicts nature as a radiant beautiful maiden. For example, in the following stanza from *Kamayani:*

> *Vibhav matawali prakriti ka aavrana vah neel,*
> *shithil hai jis par bikharata prachur mangal kheel;*
> *rashi-rashi nakhat kusum ki archana ashrant*
> *bikharti hai, tamras sundar charan ke prant!*

Vasana, 40

Here Prasad has portrayed nature as a voluptuous though graceful and beautiful lady worthy of worship. She is seen as being drunk with her beauty. Her upper raiment (*Uttariyeya*) like blue sky is negligently draped and it keeps falling off. Her beautiful feet are seen as the moon. She is being worshiped with flower like stars.

Again, the poet evokes the image of nature as a playful and mischievous pretty girl standing on the mountain peaks:

Uchcha shail shikharon par hansati, prakriti chanchala bala;
Dhawal hansati bikhrati apni phaila madhur jwala!

Karma, 50

Prasad humanizes mountains, especially the Himalayas and presents it in its different aspects/forms. The Himalayas is shown as a king-wearing moon like crown. The Himalayas is portrayed as an over ambitious man trying to reach the sky.

Chhoone ko ambar machali see barhi ja rahi satat unchayee;
Vikshat usake anga, pragat the bheeshan khadda bhayankar khai!

Rahsya, 109

The Himalayas in its over-weening desire to reach the sky jumps, falls, and hurts itself. Its cavities and gorges are wounds it gets while jumping. In the 'Vasana' canto the Himalayas is shown kissing the sky:

Dekh lo, unche shikhar ka vyom chumban vyasta

Vasana, 38

During the great deluge and destruction, the Himalayas is shown helping and rescuing the earth from getting drowned;

Vishwakalpana sa uncha vah sukh sheetal santosh nidan;
aur doobati see achala ka avalamban maniratna nidhan !

Asha, 19

Prasad humanizes the earth too. Let us take the example of the earth emerging out of the great deluge as a shy and sulking newly wed young wife, angry with her husband for not paying adequate attention and care.

Sindhu sej par dhara vadhu ab tanik sankuchit baithi see;
Pralay nisha ki halchal smriti mein maan kiye see ainthi see!

Asha, 17

The rivers and seas are also humanized. In the Ida canto, Prasad portrays the Saraswati river as a traveller who unmindful of the distractions on the way keeps steady on its course and reaches its destination. The sea waves are variously described as shouting and roaring like an angry young man, as yawning, stretching its limbs and falling off into deep slumber like a tired man, as trying to reach up to the ears of the sky and declare its love for it.

Darkness, light, shade and breeze (wind) are also humanized. Let us take the example of breeze. Prasad evokes the image of breeze as a courtier singing paeans to its king, the Himalayas:

Shila sandhiyon se takara kar pavan bhar raha tha gunjar,
us durbhedya achal dridhata ka karta charan sadrish prachar !

Asha, 19

To take another example from 'Asha' canto, the breeze is portrayed as a singer who inspired by the quiet romantic moonlit night starts singing hymns from the Sam Veda.

Dhawal manohar chandra bimba se ankit sundar swaccha nisheeth;
Jismen sheetal pawan ga raha pulakit ho pavan udgeetha !

Asha, 21

The sky is depicted as the blue-necked Shiva with bowl-like moon filled with blue poison in his hands. The stars are blue pupils of his eyes, which appear to be full of infinite compassion and peace despite having taken poison. The sky absorbs the water of the earth during summer and releases it in the rainy season to revive the earth.

Neel garal se bhara hua yah chandra kripal liye ho;
Inhi nimilit taraon mein kitni shakti piye ho!
Akhil vishwa ka visha peete ho srishti jiyegi phir se!

Karma, 51

The clouds below the Himalayan peaks are portrayed as playful young boys who wear necklace-like rainbow. They run about and play like young ones of elephants.

Neeche jaldhar daud rahe the sundar surdhanu mala pahne
kunjjar kalabh sadrishya ithlate chamkate chapla ke gahne !

Rahasya, 109

Night is depicted as a voluptuous, wanton and fickle woman who unmindful of her beauty lets the world plunder and rob it. The poet tells this young woman to drape her raiment properly so as to cover her exposed body.

Phali han sambhal le kaise chutha para tera anchal;
dekh bikharti hai mani raji ari utha besudha chanchal!
Phata hua tha neel vasan kya O yauwan ki matwali

dekh akinchan jagat lootata teri chhavi bholi-bhali !

Asha, 22

Night is also portrayed as a loving and caring mother who puts the ill-infant-like world in her lap, patting, soothing and providing solace to it:

Anchal latakati nisheethini apna jyotsana shali,
Jisaki chhaya mein sukh pawe srishti vedna vali!

Karma, 50

Important thing to note here is that the act of perceiving nature in human terms on the part of the *chhayawadi* poet makes it possible for him to interact with nature. A new domain of experience is opened up where nature is not merely acted upon, but it is seen to be acting upon human beings. Nature responds to the poet and he is seen as responding to nature. Nature stimulates and evokes emotions and feelings in the poet and the poet often expresses his emotions and feelings through nature. There is an easygoing familiarity and intimacy between the two, each sharing with and responding to the other.

In *Kamayani*, nature is shown as intimate with human beings, empathizing with them, sharing and participating in their happiness as well sorrow. Let us take the example of the description of nature at the time of union of Shraddha and Manu. Nature is not only shown as sharing and participating in the happiness of Shraddha and Manu, but it also experiences their love and passion:

Srishti hansne lagi aankhon mein khila anurag;
rag ranjit chandrika thi, uda suman parag !
. . .

Devdaru nikunja gadhar sab sudha mein snat;
sab manate ek utsav jagran ki raat !
Aa rahi thi madir bhini madhvi ki gandha,
pawan ke ghan ghire padate the bane madhu andha !
shithil alsaye padi chhaya nisha ki kant;
so rahi thi shishir kana ki sej par vishrant !

Vasana, 38, 39

The whole of nature becomes joyous with love blossoming in its eyes. The moon is full of passion and the happy and fragrant pollens fly all around. All the trees and plants including pine trees are bathed in moonlight and they all celebrate the night on which Shraddha and Manu are to unite in conjugal bliss. Flowers are spreading their sweet fragrance and the breeze is blinded by passion. The last two lines are full of sexual imagery and they show nature itself performing the sex act. Love and passion of Shraddha and Manu are transposed on to nature. We have here the image of night as a woman who, with all her passion spent in a nightlong intense and passionate love-making, now rests lazily and sensuously on the bed of morning.

Again, when Manu suffers the pangs of separation from Shraddha, the sky, the earth and the ocean not only commiserate with him, but also affect the distressed emotional state of Manu:

Ye sab sphuling hain meri is jwalamayee jalan ke
kuch shesh chinha hain kewal mere us mahamilan ke !
bulbule sindhu ke phoote nakshatra-malika tooti !
nabhayukta-kuntala dharni dikhayee deti looti !

Nature to the *chhayawadi* poet is a great repository of truths of life. He learns moral and spiritual lessons in the company of nature. Objects of nature—buds, flowers, trees, waves, rivers, mountains, clouds, dawn, dusk, etc.—reveal some truth or the other to the poet. In *Kamayani* too, nature appears as a teacher giving valuable lessons to mankind. In the 'Karma' canto we are told that man should share his happiness with others as buds after blossoming spread and share their fragrance all around:

Ye mudrit kaliyan dal mein sab saurabh bandi kar len;
Saras na ho to makrand bindu se khulkar to ye mar len!

Karma, 55

If the buds gather up and fold all fragrance in themselves and do not share it with the world, then they will remain undeveloped and unblossomed and they will die and fall on the ground without having developed and blossomed fully. Similarly, if a man does not share his happiness with others, he will also meet a sad demise, his personality without having been developed fully.

In the 'Shraddha' canto Shraddha explains the complementary relationship between happiness and sorrow by drawing parallel with night and day and emphasizes the need for man to remain unmoved by both.

Dukh ki pichhli rajani beech vikasta sukh ka nawal prabhat;
Ek parda yah jheena neel chipaye hai jisme sukh gaat!

Shraddha, 26

As the golden morning follows the dark night, happiness follows sorrow. As morning remains hidden behind the darkness of night, happiness also remains hidden behind the darkness of sorrow and happiness always comes at the opportune and appointed time. As night and morning are two aspects of the day, sorrow and happiness are also two essential aspects of life.

The *chhayawadi* poet is a nature mystic who sees the Divine Spirit revealed through nature. He perceives the presence of the Divine Spirit even in the smallest particles of nature. In *Kamayani* too Prasad sees in nature the presence of Shiva who governs the entire creation:

Vishwadev, savita ya pusha, som, marut, chanchal pavman
Varuna aadi sab ghoom rahe hain, kiske shasan mein amlan?
Kiska tha bhru-bhang pralay-sa jismen yah sab vikal rahen,
Arey ! prakriti ke shakti-chinha the phir bhi kitne nibal rahe !

Asha, 17-18

Everything in nature points to the existence of the immanent and transcendent Shiva, but all attempts to define and describe him fails because he is formless, infinite, indescribable and inscrutable:

Mahaneel is param vyom mein, aantariksha mein jyotirman
Grah, nakshatra aur vidyutkana kiska karte ai sandhan?

Sir neecha kar kiski satta sab karte sweekar yahan;
Sada maun ho pravachan karte jiska, vah astitva kahan?
hey ananta ramaniya ! kaun tum? yah main kaise kah sakata?
Kaise ho? Kya ho? Iska to bhar vichar na sah sakta
He Virat ! He Vishwadev ! Tum kuch ho aisa hota bhan!

Asha, 18

It is the presence of the Divine Spirit in everything that makes the poet see unity in diversity, the universal in particular:

Chetan samudra mein jeewan lahron sa bikhar para hai;
Kucha chhap vyaktigat, apna nirmit aakar khada hai !

Asha, 18

The Divine Spirit is like the ocean and all entities living and non-living are its waves. Each wave has a particular temporal identity insofar as it looks different from other waves and yet all these waves are created out of the water of the ocean and they remain inseparable from it. So man and nature though different in appearance are essentially one and the same. Manu says:

Main ko meri chetanata sabko hi sparsh kiye si;
Sab bhinna paristhitiyon ki hai madak ghoont piye si!

Anand, 118

He feels that it is his consciousness, which pervades everything. And it is this awareness that makes Manu seek identification with nature by telling Mount Kailash that they are inseparable and one:

Manu ne kucha kucha muskayakar Kailash or dikhlaya;
Bole "dekho ki yahan par koi bhi nahin paraya!
Ham anya na aur kutumbi ham kewal ek hamin hain;
Tum sab mere avyav ho jismen kucha nahin kami hai"!

Anand, 118

He obliterates all distinctions between self and other, man and nature and experiences oneness with the world/nature. This universe becomes a nest where all beings exist in harmony and peace:

Sab bhedwav bhulwakar dukh-sukh ko drishya banata;
Manav kah re! 'Yah main hoon' Vah vishwa needa ban jata!

Anand, 119

Thus, we see that nature, for Prasad is not an object, lifeless and inorganic matter, but a presence and a power, a motion and a spirit, not something to be used and consumed, but always a guide leading beyond it. Nature itself leads him beyond nature to the infinite and inscrutable presence, which permeates and pervades all. This process and act of going beyond nature also involves his becoming one with nature. Man is not an intruder into the pasture of nature but a devotee who searches for the essential moral and spiritual values in nature, which makes life worth living, full of peace, happiness and harmony.

Reference

Prasad, Jai Shankar (1935) [1992] *Kamayani*, Delhi, Arun Prakashan.

19

Asaadhya Veena: A Poetic Symbol of Deep Ecological Self-realization

Sujata Chaturvedi

The theory of deep ecology in environmental ethics deals with the ecocentric view, wherein self-interests of humans are integrated with nature. An aspect of deep ecology is self-realization, which allows us to contemplate and adopt a more mature, non-egoistic and sympathetic outlook towards the entire human as well as non-human world. The philosophy of self-realization has been a part of Indian philosophy since ages and has been repeatedly expressed in Indian literature and religious scriptures also. An attempt has been made through this paper to relate the theory of self-realization in a multifaceted and touching poem *Asaadhya Veena* written by the renowned Hindi poet 'Agyeya' (Sachhidanand Hiranand Vatsyayan 'Agyeya'). Various aspects, shades of meanings of the beautiful poem have been discussed and the importance of self-realization or 'Atma-Shodhan' and dilution of narrow ego has been stressed upon. As an interdisciplinary study, it opens new vistas of research into the basic philosophy of literary works and tries to highlight the important relation between contemporary philosophical and literary trends and their deep extensive impact on the entire society.

Human life and environment are interdependent and need each other for their sustenance and growth. Ecology comprehensively explains the deep-rooted relationship between living organisms and their environment. Hence, in a moral sense, it becomes our ethical duty to protect and develop our environment. The same ecosystem that sustains us also expects us to help in maintaining its natural balance. Here steps in the most recent and very important concept of environmental ethics. All over the world, there is now a growing concern for ethical norms in all spheres of human activities. Environmental ethics concern the value system of societies, the value system that has brought the state of environment to the present situation in which there is exploitation, not only of nature, but also of some societies by the others.

There is one view which questions the need for any change in values and attitude of people *vis-à-vis* environment. No change in attitude or value system is required. This is termed as the 'anthropocentric theory' in which human beings are at the centre, they stand apart from nature and their destiny is to master nature. Anthropocentric view, therefore, is in contrast to ecological thinking. On the other hand, according to eco-centric view, nature and human beings have equal status and therefore, equal rights. Those who subscribe to eco-centricism believe that human self-interests and those of nature are same.

From this perspective of a life-centred theory, humans have moral obligations that are owed to wild plants and animals themselves as members of the earth's biotic community.

According to Paul W. Taylor:

> We are morally bound (other things being equal) to protect or promote their good for their sake. Our duties to respect the integrity of natural ecosystems, to preserve endangered species, and to avoid environmental pollution stem from the fact that these are ways in which we can help make it possible for wild species populations to achieve and maintain a healthy existence in a natural state.

The eco-centric or bio-centric view, hence, considers humans as members of earth's community of life. The earth's natural ecosystems in totality are seen as a complex web of interconnected elements with sound biological functioning of the others.

It is with this bio-centric view of environmental ethics that the term 'deep ecology' is connected. Arne Naess coined this term 'deep ecology' in his 1973 article *The Shallow and the Deep, Long-Range Ecology Movements*. Bill Deyal and George Sessions have explained this concept as: "The essence of deep ecology is to keep asking more searching questions about human life, society and nature as in the western philosophical tradition of Socrates, thus deep ecology goes beyond the so-called factual scientific level to the level of self and earth wisdom."[2] Arne Naess has developed two ultimate norms of deep ecological consciousness. They are arrived at by the deep questioning process and reveal the importance of moving to the philosophical and religious levels of wisdom. These ultimate norms are 'self-realization' and 'biocentric equality'.

Self-realization process begins when an individual's enlarged egoism replaces a more mature and an all-encompassing attitude. An isolated ego strives for narrow individualistic goals; spiritual growth begins when we not only identify ourselves with other humans but also with the non-human world. Self-realization makes us see beyond our narrow contemporary cultural assumptions and values. It allows us to introspect and question our motives and actions. This meditative deep questioning process through regular practice becomes a habit that transforms an individualistic attitude to a mature, serious, non-egoistic and sympathetic outlook towards the entire humanity as well as the non-human world. Increased self-realization implies broadening and deepening of the self. Hence, self-realization ultimately liberates the *atma* and helps it attain *moksha*, i.e., liberation of the self and attainment of closeness to the ultimate power or *Paramatma*. This process ensures the unity of *atma* with *Paramatma*, which is the ultimate spiritual goal of human life.

Arne Naess, the Norwegian philosopher, whose ideas are the main inspiration of the deep ecology movement, puts forth the view that by identifying with greater wholes, we partake in the creation and maintenance of this whole. In his words, "The intensity of identification with other life depends upon milieu, culture and economic conditions. The ecosophical outlook is developed through identification so deep that one's own, the personal ego or the organism no longer adequately delimits self. One experiences oneself to be genuine part of all life".[3] Naess further makes it clear by stating that when solidarity and loyalty are powerfully embedded in identification, they are not experienced as moral demands but they come by themselves naturally. From the process of identification stems unity, and unity ultimately results in wholeness. This identification makes one realize the importance of interconnectedness of everything and that our egos are mere fragments, which have an extremely limited power within the whole. Hence, through the deep power of identification our egos develop into selves of greater dimension and depth.

A very fine example of this process of identification, oneness and self-realization is the long poem—*Asaadhya Veena* composed by the renowned Hindi poet Sachhidanand Hirananda Vatsyayan 'Agyeya'. Agyeya is one of the most important and creative poets and authors of modern Hindi literature. His contribution to literature is deeply bound with philosophy, history and intense creativity along with great intimacy with folklore. This poem *Asaadhya Veena* is based on a Japanese story—"Taming of the Harp", where a harp or 'veena' is made from a huge tree called 'Kiri'. The harp is made by a magician and gifted to the king of China. The king is unable to find anyone who could play the harp. In the end, Peevo was able to play the harp and that was because he diluted his ego and identified himself completely with the harp. Agyeya has used this Japanese tale in the form of a myth in *Asaadhya Veena*. The beauty of a myth lies in its constant renewable form. It crosses all boundaries of society, culture and civilization and expresses itself in accordance with new trends in different cultures. The creativity and *aastha* (deep belief) of the poet has been explicitly told in *Asaadhya Veena*.

The poet begins this poem in a narrative form where an ancient veena gifted to a king is kept in his Darbar or court and nobody has been able to produce sound or *swar* from that veena. Hence, it got the name *Asaadhya Veena*, i.e., one, which cannot be attained or controlled or reached:

> *Mere haar gaye sab jaane-maane kalaavant,*
> *Sabki Vidya ho gayee akaarath, darp choor*
> *Koi gyani gooni aaj tak ise na Saadh Saka*
> *Ab yeh Asaadhya Veena hi khyat ho gayee.*[4]

The king has also lost all hopes of hearing any sound from that veena, when one day a great sage Priyamvad arrives in his Darbar. The veena is offered to him to play. The sage first bows his head on the veena and then brings it close to himself and keeps it on his lap so that his head touches the veena:

> *Vaadya uthaa saadhak ne gode rakh liya Dheere-dheere jhuk us par,*
> *taaron par mastak tek diya.*[5]

It is from here that Priyamvad's journey into inner-self begins. He moves from the plane of the outer world into the plane of meditation and deep realization. To attain *swar* or sound from veena is to understand and know the essence of veena:

> *Par us spandit sannate mein*
> *Maun Priyamvad Saadh raha tha Veena-*
> *Nahin, Swayam apne ko shodh raha tha.*
> *Saghan nivir mein wah apne ko*
> *Saunp raha tha usi kiriti-Taru ko.*[6]

To understand that essence Priyamvad travels internally to the source of birth of that veena. His internal self meditates and moves to the deep and huge forests of Uttarakhand, at the foothills of Himalayas. In this forest stands tall and broad, the Kiriti Taru or the tree from which the veena was shaped. Priyamvad tries to identify himself with the Kiriti Taru and tries to reach the tree through deep meditation and extensive feeling of oneness with the environs of the tree:

O Vishal Taru!

O deerghkaay!
O Poore Jhaarkhand ke agraj,
Taat, jSakhaa, guru, ashray,
Traata, Mahachchhai.

Mein tujhe sunoon
Dekhoon, dhyaoon
Animesh, Stabdh, Sanyat, Sanyut, nirvaak.[7]

After this the poet, through Priyamvad, experiences the entire non-human environment of the great tree—all, that is in, on and around the Kiriti Taru. This includes the lions, tigers, bears, deer, small insects, birds, leaves, flowers; in short, everything that is part of the ecosystem. The underlying message behind this is also that the Kiriti Taru feels great happiness and contentment in supporting all these life systems growing on and around it. It stands as a pillar of direction for the humans, showing the deep feeling of fulfilment in caring for others. "The greater our comprehension of our togetherness with other beings, the greater the identification, and the greater care we will take."[8] This also results in the delight in the well-being of others and sorrow when harm befalls them:

Wahi Kiriti Taru

Uske kaanon mein him-shikhar rahasya kahaakarte the apne,
Kandhon par badal sote the

Kotar mein bhaloo baste the
Kehari uske valkal se kandhe kujlaane aate the.[9]

This is how Priyamvad feels oneness with the Kiriti Taru (Taru meaning tree) and all the lives supported by it. This feeling of oneness is possible only when the individual ego or *aham* is totally dissolved. The poet makes Priyamvad travel through various seasons and experience all the climatic changes that the tree bears and the effects of the varying moods of nature on the tree. This meditative journey takes him from summer to rainy and then to winter season.

Haan, majhe smaran hai:
Badli-kondh-pattiyon par varsha-boondon ki pat-pat.
Ghani raat mein mahue ka chupchap tapakana.[10]

Priyamvad experiences even the smallest sound of a cricket and slightest movement of a dew drop and on the other hand the sound of horse's hoof and festive songs in a remote hilly village:

Parvati gaon ke utsav-dholak ki thaap

Os-boond ki dharkan itni komal, taral, ki
Jharte-jhartemano
Harsingar ka phool ban gayee.[11]

All these sounds add up to form a great canvas of experiences that the Kiriti Taru has had in its life. Since the veena is basically a part of that tree, so all these sounds form a part of the essence of veena also. In order to draw music from the veena, it is essential to know the background, the basic formative structure and the environment of the veena, and that is what Priyamvad does when he explores his inner self and journeys towards the Kiriti Taru:

Sunta hoon main
Par har swar kampan leta hai mujko mujse sokh

Mujhe smaran hai-
Par mujko main bhool gaya hoon.[12]

This whole process of realization occurs through identification. Priyamvad identifies himself with the veena and the Kiriti Taru and feels one with them. The poet Agyeya renders the basic concept of deep ecology beautifully in this poem. It is through identification and self-realization that Priyamvad finally strikes the strings of the veena and divine music emanates from it. Again, at this point, the poet moves to a higher plane, an all-encompassing bond, where the sound of veena is one but each individual interprets it according to his/her mental status and environment. The king hears the sound of veena as the sound of victory and peace; the queen interprets it in the form of variety of jewellery items and then the feeling of all pervading love which conquers the lust for worldly assets:

Mekhla-kinkini
Sab andhkar ke kan hain ye-aalok ed hai
Pyar ananya!

Rani
Us ek pyar ko saadhegee.[13]

Similarly, all the *Durbaris* (courtesans) interpret the sounds of veena as per their innermost and long-awaited wish. They all experience a deep feeling of satisfaction and contentment. Life seems to have become simpler and easier after listening to the veena, as each one develops a bond with the other and forgets the confrontations and individualistic narrow egos of the day-to-day life. This feeling of contentment is attained only after one reaches the stage of self-realization. Hence, Priyamvad not only went through that process of self-realization, but he permeated that experience through the music of veena to one and all:

Sab doobe, tire, jhipe, jaage-
ho rahe vashamvad, stabdh:
Eyataa sabki alag-alag jaagi,
Sandheet hui,
Pa gayee vilaya.[14]

There are two points here to which we should pay attention. First, is 'music' and the deep inner strength of music. Music in itself is *sadhna*, i.e., meditation. In order to reach the platform of meditation, one has to go through the process of self-realization. Hence, a person who is able to produce music will definitely be able to connect with the whole world, the entire life—human and non-human, and create a whole gamut of emotions and feelings through the binding power of music. Indeed, music has a very fine power of healing and binding together. In today's world of discontentment and disharmony, music is one force, which can bind humanity and the entire ecology together so that they move together in harmony, balance and equilibrium. The poet Agyeya has put forward a very strong source of maintaining the harmony of ecology and that is the indefatigable power of music. Music, hence, heals us from inside and binds us from outside and therefore increases our realization of the self and deepens the awareness of our environs and ecosystems:

> *Avtarit hua sangeet*
> *Swayambhoo*
> *Jisme sota hai akhand*
> *Brahma ka maun*
> *Ashesh Prabhamay.*[15]

The second point to be highlighted here is the manner of interpretation in which various courtesans define the music of veena. Each one finds a meaning in the music with reference to their individual sphere of work and living. This interpretation on one hand expresses the feeling of satiation of each one's deepest desires and on the other hand points to an important aspect of *sadhna*, which is meditating while performing day-to-day duties, of *grihastha* (family life). This *sahaj sadhna* (as propagated by various religions, Kabir, Buddha and others) holds forth the importance of our earthly duties and the fact that we are responsible for our actions and our environment. Through the process of sadhna or meditation we can cleanse our inner self daily and thereby our thoughts and hence actions towards others also. It is towards this kind of meditation that Agyeya wishes to point through the music of veena.

In the entire poem *Assaadhya Veena*, the poet has given a strong message of the theory of self-realization in the *Deep Ecology Theory*. This forms a basic part of the Indian philosophy and has been emphasized and propagated time and again by eminent philosophers, religious leaders and litterateurs, etc. Agyeya wishes to put forward the fact that as long as a person's narrow ego or *aham* overshadows his being, the veena remains *asaadhya* (unconquerable). But, when one meditates and dilutes one's ego (bows the head in front of the great power) and realizes the strength and expanse of *Parmatma*, it is then that the *asaadhya* becomes *saadhya* and divine music can be heard and assimilated. This becomes equivalent to *moksha* or *mukti*:

> *Main nahin, nahin! Main kahin nahin!*
> *Mujhe kshama kar-bhool akinchanta ko meri*
> *Mujhe oat de-dhank le-chaa le-*
> *O Sharanya!*
> *Mere goongepan ko tere soye swar-sagar ka*
> *Jwar duba le!*[16]

There are traces of Buddhist philosophy in *Asaadhya Veena* at some places. The Buddhist idea of 'Nirvana' is quite similar to what Agyeya has put forth in his poetry. Dr. S. Radhakrishnan has explained that "Nirvana, which is the consummation of the spiritual struggle, is a positive blessedness. It is the goal of perfection and not the abyss of annihilation. Through the destruction of all that is individual in us, we enter into communion with the whole universe and become an integral part of the great purpose. Perfection is then the sense of oneness with all that is, has ever been and can ever be.... It is a kind of existence devoid of egoity, a timeless existence, full of confidence, peace, calm, bliss, happiness, delicacy, purity, freshness."[17] In Buddhism, it is further explained that in Nirvana, which is compared to deep sleep, the soul loses its individuality and lapses into the objective whole. The wind of ignorance blows over it and stirs its equable flow, causing vibrations in the ocean of existence. The sleeping soul is wakened and its calm unfettered course is arrested. It wakes up, thinks, builds individuality and isolates itself from the stream of being and resuming the uninterrupted flow. "Nirvana.... is becoming one with the eternal reality.... It is the fading of the star in the brilliant rise of the sun or the melting of the white cloud in the summer air."[18] Hence, the false individuality disappears and only the true being remains. This, again, is possible only through the process of self-realization and identification.

In *Asaadhya Veena* the poet has left open space for immense imagination and deep philosophizing process. Priyamvad shows us the way to meditation and then realization of the self. We can also understand the essence of the poem by interpreting the veena as one's own self or *atma* and the Kiriti Taru as the divine *parmatma*. Just as the veena is grafted from the Taru, so are we a part of the supreme power (or God) and so is *atma* a part of *parmatma* (*Parma* meaning supreme and *atma* meaning self). It is through the process of dispersion and dilution of our ego and then the journey of self-realization that the veena can emanate music. So is the case with *atma*, that it can emanate divine music only when it goes through the process of self-realization. This divine music has the power of healing and binding the whole ecology and hence *atma* feels the deep oneness with the whole *srishti* (universe).

> *Shreya nahin kuch mere:*
> *Manin to doob gaya the swayam shoonya mein*
> *Veena ke madhyam se apne ko maine*
> *Sab kuchh ko saunp diya tha.*[19]

Then again, there comes another underlying layer of message to be discovered through this immensely beautiful poem, that is, the "message of harmony with nature". Man was created as a part of this nature and in order to maintain the ecological balance, man is expected to remain a part of nature and develop and grow in tune with the nature. According to the ecocentric view of environmental ethics also, humans are members of the earth's community of life and should contribute in the growth and development of life and not indulge in undue interference and exploitation of non-human spheres. It is of utmost importance that in today's growing awareness towards safeguarding the environment and ecology, one should pay attention to the fact that nature has its own lifecycle and if humans respect that pattern of cycle and remain a natural part of it, then gradually the equilibrium of nature can be restored. This is possible, again, only when the humans learn to deny their superiority complex over the mighty nature. This can be achieved only through self-realization or *atma-shodhan*, Agyeya has repeatedly emphasized this point in his various writing that nature and man have to live in harmony and that man should dissolve his false sense of narrow, constricting ego or *aham*, only then liberation or *mukti/moksha* can be achieved.

There is yet another facet to the poem that can be explored and that is "the process of making of poetry", Agyeya has also tried to explain the inner journey of a poet when he is giving birth to a poem. A poem is the expression of one's deepest self, a process of realization of one's minutest, complex thoughts and a medium of expression of one's cherished desires as also it is an effective medium of awakening the society also. When a writer/poet embarks on a journey of writing his thoughts, it is a journey of *sadhna* or meditation as he looks into his inner self, innermost thoughts and ideas and realizes his own 'self'. Only then is he capable of giving birth to his new creation. A poet has to meditate deep into the environs of his subject, just as Priyamvad did while meditating and thinking of forests of Uttarakhand and Kiriti Taru. When the poet feels *tadatmya* (feeling of oneness) with the nature and the subject, it is only then that the divine music flows from the veena, i.e., poetry is born from the heart of the poet. Hence, it is again the process of self-realization through which the thought-invoking writings or poetry is born, which in turn gives a meaningful direction and message to the whole human society.

Therefore, *Asaadhya Veena* is basically a symbol of deep ecological self-realization in poetic form. It is a much needed and contemporary guiding light towards conservation and protection of environment. More than that *Asaadhya Veena* embarks on a fruitful journey of meditation, harmony, balance and realization of a mature self. It moves beyond the boundaries of humans and steps into the unbounded world of entire ecological system and spreads the powerful message of love and harmony. The poem not only enlightens us enough to make our thinking faculties spread their wings, but also opens the vistas of imagination, binding it together with philosophy of life itself. It puts forward a philosophical base of meditation on which the spiritualistic form of poetry oriented with environmental ethics is built. *Asaadhya Veena* is truly in tune with the times and should be taken as a guiding light towards the goal of self-realization. The poem has in-built seriousness, softness, sympathy and somberness towards the very important issue of environmental ethics. Thus, Agyeya has certainly exposed a deep sense of environmental harmony and importance of *atma-shodhan* (self-realization) and necessity of dilution of *aham* (ego) through this very meaningful, philosophic rendering of literary thoughts. It is apt to end the paper with Agyeya's philosophic thoughts:

> *Suna aapne jo wah mera nahin,*
> *na Veena ka tha:*
> *wah to sab kuchh ki tathtaa thi-*
> *Mahashunya*
> *Wah mahamaun*
> *Avibhajya, Anapt, Adrawit, Aprameya*
> *Jo Shabdaheen Sab mein gaata hai.*[20]

Notes

1. Paul W. Taylor, "The Ethics of Respect for Nature", in Donald Van DeVeer and Christine Pierce, *The Environmental Ethics and Policy Book*, Wadsworth Publishing Company.

2. Bill Deval and George Sessions, "Deep Ecology" in Donald Van DeVeer and Christine Pierce, *The Environmental Ethics and Policy Book*, Wadsworth Publishing Company.

3. Arne Naess, "Identification, Oneness, Wholeness and Self-realization" in John Benson, *Environmental Ethics*, Routledge , London and New York, p. 245.

4. Agyeya, *Asaadhya Veena*, p. 34.

5. Ibid., p. 35.

6. Ibid., p. 36.

7. Ibid., p. 37.
8. Arne Naess, "Identification, Oneness, Wholeness and Self-realization" in *Environmental Ethics*, John Benson (2000), Routledge, p. 246.
9. Agyeya, *Asaadhya Veena*, p. 34.
10. Ibid., p. 39.
11. Ibid., p. 39.
12. Ibid., p. 41.
13. Ibid., p. 44.
14. Ibid., p. 45.
15. Ibid., p. 434.
16. Ibid., p. 42.
17. Dr. S. Radha Krishanan (1929), *Indian Philosophy*, Vol. I, Oxford University Press, p. 448.
18. Ibid., p. 450
19. Ibid., p. 46.
20. Ibid., pp. 46-47.

References

Agyeya, *Asaadhya Veena* in *Aangan ke Paar Dwar*, 1961.

Benson, John, *Environmental Ethics*, Routledge, London and New York, 2000.

Buddhist Suttas, Sacred Books of the East, Vol. XI.

Chaturvedi, Ram Swarup, *Agyeya aur Aadhunik Rachna ki Samasya*, Bhartiya Gyanpeeth, 1968.

Johnson, Lawrence E., *A Morally Deep World*, Cambridge University Press, 1991.

Mill, J.S., "The Principles of Political Economy' in *Collected Works*, Vol. 3, University of Toronto Press, 1963.

Mishra, Ram Darash, *Aadhunik Hindi Kavita: Sarjanatmak Sandarbh*, 1986.

Mishra, Vidyaniwas (ed.), *Aaj ke Lokpriya Kavi: Agyeya*, Rajpal and Sons, Delhi.

Naess, Arne, *Economy, Community and Life Style*, Cambridge University Press, 1989.

Radhakrishanan, S., *Indian Philosophy*, Vol. I, Oxford University Press, 1929.

Shah, Ramesh Chandra (ed.), *Agyeya aur Asaadhya Veena*, National Publishing House.

Taylor, P.W., "The Ethics of Respect for Nature", *Environmental Ethics* 3 (3), 1981.

Tiwari, Vishwanath Prasad (ed.), *Agyeya aur unka Sahitya*, National Publishing House, 1972.

Van DeVeer, Donald and Christine Pierce, *The Environmental Ethics and Policy Book* (2nd ed.), Wadsworth Publishing Company.

Zimmerman, M.E. (ed.), *Environmental Philosophy*, Prentice Hall, 1993.

IV

Science, Society and Environment

20

Energy Resources, Impact on Environment and Ethical Problems

P.K. Rath

Energy is one of the primary necessities for the creation, existence and sustenance of the physical as well as biological world. William Blake said: "Energy is eternal delight." In physical sciences, energy is defined as the capacity for doing work. Energy is both of these and more. According to Emmy Noether, symmetries in nature are associated with conserved quantities and vice versa. Certain physical quantities are conserved due to the space-time symmetries of our physical world. Momentum and angular momentum are conserved due to the homogeneity and isotropy of physical space. Similarly, the total energy is conserved, as time is homogeneous. Hence, energy can neither be created nor destroyed. It can only be transformed from one form to the other and in particular from latent to usable form.

According to Albert Einstein, all forms of matter in the physical world are sources of energy. However, it is not profitable to derive energy from any form of matter. The first and second laws of thermodynamics have far reaching implications for the physical processes to be used in deriving energy from any source. The first law simply states that energy will remain constant during such conversion. However, the second law is quite important in the sense that it decides the practical feasibility of any process to occur in nature. The Clausius statement of the second law of thermodynamics is that no process is possible whose end result is the transfer of heat from a colder to a hotter body. In other words, heat of its own accord always flows from higher to lower temperature. The Kevin statement of the second law of thermodynamics is that no process is possible whose end result is the complete conversion of heat into work. All the heat supplied to any system cannot be entirely converted into useful work, since heat has to be always rejected to the outside reservoir. Thus, given the second law of thermodynamics, thermal pollution is an inevitable fact of nature. It is neither intentional nor avoidable. Improving the efficiency of the conversion process can only minimize it.

To examine the nature of other kinds of pollution, a working definition is needed. In November 1965, the environmental pollution panel, of the Science Advisory Committee of the president of USA, in its report, "Restoring the Quality of Our Environment", adopted the following definition of pollution:

> Environmental pollution is the unfavourable alternation of our surroundings, wholly or largely as a by-product of man's action through direct or indirect effects of changes in energy patterns, radiation levels, chemical or physical constitution and abundance of organisms. These changes may affect man directly or though his supplies of water and of agriculture and other biological products, his physical objects or possessions or his opportunities for recreation and appreciation of nature.

The production of pollutants come as a by-product of man's actions—they are the residues of things men make, use and throw away—their cans and bottles, metals and plastic caps, waste rock and mill tailings, pesticides and herbicides, automobile exhausts and industrial discharges. They are concomitants of a technological society with a high standard of living. They increase, both because of population increase and because of an increasing expectation of higher living standards. Hence, pollution can be regarded as primarily a result of human behaviour that largely disregards consequences, one that unwisely uses and disregards resources. However, pollutants also include the natural byproducts of human metabolic activity and that of organisms on which humans depend for food, their biological waste products and farm animal excreta. Hence, before we discuss the ethical problems associated with environmental pollution or misplaced resources of nature in the context of energy resources, we briefly discuss the origin of energy resources and their impact on environment.

Energy Resources

Energy resources can be classified according to their availability, origin and use. With respect to the availability, energy resources can be classified as captive or renewable. The energy resources are divided into those of solar, lunar, cosmic and geological origin (Table 1). They are classified as conventional or non-conventional or alternative according to their use.

Table 1
Energy Resources of Different Origin

Solar origin	Lunar origin	Cosmic origin	Geological origin
Fossil fuel	Tidal	Fission	Geothermal
		Fusion	
Hydroelectric			
Solar energy			
Deep ocean currents			
Wind power			
Biomass			

In the following, we briefly discuss various energy sources according to the classification based on use:

Conventional Energy Resources

The conventional energy resources are fossil fuels, hydroelectricity and nuclear power.

Fossil Fuels

Fossil fuels, namely, coal, oil and gas, are organic materials, which have been slowly broken down by biological and geological processes. Oil and natural gas along with coal are termed minerals, as they are a part of the make up of the earth's crust and are hydrocarbons. Coal consists of plant remains whereas oil and gas are mainly composed of animal matter. Hence, fossil fuels are a part of the carbon cycle that is basic to life on earth. There is probably about 50 times as much carbon locked up in fossil fuels as there is in all living matter on earth.

(a) *Coal*: Coal was the first fossil fuel to be exploited on a large scale. It made possible the industrial revolution, which in turn benefitted the coal industry by providing it with superior technology, thereby enabling coal to be mined at greater depths. Coal is a sedimentary rock consisting of carbon, water and volatile gases with small amount of mineral impurities, which produce ash when the coal is burned. The main mineral impurities are clays, chloride and sulphides. When sulfides mix with air during combustion, what results is sulfur dioxide, which is an important cause of atmospheric pollution.

There are three different types of coal, namely: (i) lignite, (ii) bituminous, and (iii) anthracite, depending on the presence of different proportions of the constituents. The constituents of different types of coal are summarized briefly in the below:

(i) *Lignite*: Lignite, which is commonly known as brown coal, contains a higher proportion (about 43%) of water. Hence, the calorific value of lignite coal is low.

(ii) *Bituminous*: Bituminous, which is the household coal, contains only 3 per cent of water and a higher proportion of 32 per cent of volatile. Hence, bituminous is more readily inflammable than other types.

(iii) *Anthracite*: It contains about 96 per cent of carbon and hardly any water or volatile. It has a higher calorific value but is less readily inflammable than bituminous.

It is almost impossible to know exactly how much coal is in the ground. However, a reasonable estimate is that there are about 6,00,000 million tons (about 778.90 million tons in India) of coal that could be dug up using presently available mining techniques. With the present day mining rate of about 2,000 million tons per year and with some additional increase in mining rate, the known coal resources will be exhausted in about two centuries.

(b) *Oil and natural gas*: Natural petroleum or crude oil is a liquid ranging in colour from yellow to black including red, brown and dark green. It is a mixture of many hydrocarbon compounds and its viscosity ranges from very fluid to highly viscous.

Oil shale, which is a sedimentary rock containing solid combustible organic matter in a mineral matrix, is a potentially significant source of energy. Tar sand, which is a loose to consolidated sandstone or porous carbonate rock, impregnated with a heavy asphaletic crude oil too viscous to be produced by conventional means, is another kind of fossil fuel. Natural gas contains smaller and lighter hydrocarbon molecules and is colourless.

Hydropower

Hydropower, which is the energy of the fast flowing or falling water was well known to Greeks and Romans during the first century BC. The water mills were used for grinding corn. In the medieval period, Britain had as many as 5,624 water mills, which were used not only for grinding corn but for operating bellows and hammers in forging iron, for sharpening tools and weapons, for textile manufacture, for tanning and even for pumping water from mines. The power output from hydropower varied from few kilowatts to 50 kilowatts. These water wheels are not appropriate for electricity generation because of their slow motion.

The first effective hydraulic or water turbines began to appear in the second half of the 19th century. Most of the water turbines are used to generate hydroelectricity by exploiting the run-off water from mountain areas, which is stored behind a dam. Hydroelectricity, besides being a renewable source of energy, is cheap and pollution free.

Nuclear Energy

The most unfortunate aspect of nuclear energy is its association with the mass destruction of Hiroshima and Nagasaki in August 1945 by nuclear bombs. The nuclear bomb is the uncontrolled fission or fusion reaction whereas nuclear energy is produced through controlled reactions. The atomic nucleus is the storehouse of nuclear energy. Practically most of the mass of the Universe is contained in the atomic nucleus, which in turn consists of nucleons, that is protons and neutrons. Hence, according to Einstein's $E = Mc^2$, all the energy of the Universe is contained inside the nucleus. However, it is not very profitable to extract energy from all the nuclei. The process to be utilized to extract nuclear energy is decided by the binding energy, which is responsible for the binding of nucleons inside the nucleus. In other words, it is necessary to supply energy equal to the binding energy of the nucleus to make all nucleons free. Hence in a reaction, if a system moves from a less bound state to a more bound one, then energy is released. It is quite clear from the nature of binding energy data that there are two alternatives, namely fission of heavy nuclei and fusion of light nuclei to extract energy from the nucleus. We present a brief description of the fission and fusion processes:

(a) *Nuclear fission*: The nuclear materials, which can be used in fission reactions, are divided into fissile and fertile materials. The fissile materials are those materials, which can readily undergo fission with thermal neutrons. ^{235}U is the only naturally occurring fissile material with (0.71%) abundance. The other two fissile materials are ^{233}U and ^{239}Pu, which do not occur in nature but can be produced from ^{232}Th and ^{238}U through the interaction of neutrons in breeding reactions. ^{232}Th and ^{238}U are the fertile materials from which fissile material can be produced.

The first fission reaction observed by Otto Hahn and Fritz Strassmann in 1938 was:

$$n + {}^{235}_{92}U \rightarrow {}^{141}_{56}Ba + {}^{92}_{36}Kr + 3n$$

In Table 2, we have displayed the mass difference between the reactants and products, which is converted into energy in the fission reaction. The mass difference 0.215 amu converted into energy is equal to about 200 MeV. There are about 10^{22} atoms of ^{235}U in 1 kg of naturally occurring uranium. Hence the total energy released is equal to nearly 10^{11} Joules, which is a large amount. Thus, the energy released by 1 kg of uranium is equivalent to the energy delivered by the combustion of 3,000 tons of coal.

Table 2

The fission energy is due to the mass difference of reactants and products

Initial nucleus	Initial mass (amu)	Final nucleus	Final mass (amu)
$^{235}_{92}U$	235.0439	$^{141}_{56}Ba$	140.9139
Neutron	1.0087	$^{92}_{36}Kr$	91.8973
		3n	3.0261
Total =	236.0526		235.8373

(b) *Nuclear fusion*: Besides, nuclear fission of heavy nuclei, the fusion of light nuclei is also an energy releasing process. As an example, consider the following reaction

$$^2_1 H + ^2_1 H \rightarrow ^3_2 He + n$$

In Table 3, we have displayed the mass difference between the reactants and products in the above fusion reaction, which is converted into energy. Thus, the mass difference between the reactants and products 0.004 amu converted into energy is equal to 6×10^{-13} Joule. In two kg of deuterium, there are about 6×10^{26} atoms. Hence, the total energy release will be 4×10^{13} Joules.

Table 3

The fission energy is due to the mass difference of reactants and products

Initial nucleus	Initial mass (amu)	Final nucleus	Final mass (amu)
$^2_1 H$	2.015	$^3_2 He$	3.017
$^2_1 H$	2.015	N	1.009
Total =	4.030		4.026

Thus, in the fusion reaction, much more energy is released than the fission reaction. However, it is extremely difficult to initiate the fusion reaction. The fission reaction can be started with any stray neutron. But for the fission reaction to be initiated, the Coulomb repulsion between deutron atoms has to be overcome by a temperature of the order of 10^{10} ^{0}C, which is the temperature in the interior of sun and stars. Hence, fusion is also known as thermo-nuclear reaction. In uncontrolled fusion reaction that is in a hydrogen bomb, such temperatures are momentarily obtained in an atomic explosion of uranium or plutonium. Controlled fusion reaction has not been achieved so far. The most promising projects Tokmak (toroidal magnetic chamber), JET (Joint European Torus) and laser fusion are in progress. In 1955, eminent Indian scientist H.J. Bhabha in his presidential speech of the first international conference on peaceful uses of atomic energy referred to controlled fusion reaction as follows:

> We are exhausting the reserves, which have been built up by nature over a long period of time in a few centuries, in a flash of geological time. Our presently known reserves of coal and oil are insufficient to enable the under developed countries of the world, which contain a major part of the population to attain and maintain for long a standard of living equal to that of the industrially most advanced countries.... The acquisition by man of the knowledge of how to release and use atomic energy must be recognized as the third great epoch in human history.... I venture to predict that a method will be found for liberating fusion energy in a controlled manner within next two decades. When this happens, the energy problems of the world will truly have been solved, for the fuel will be plentiful as the heavy hydrogen is available in the oceans.

Non-Conventional Energy Resources

Solar Energy

The solar power is an ancient idea, very much familiar to Greeks and Romans. In 213 B.C., Archimedes used reflections from hundreds of polished mirrors to counter a Roman naval attack. With the development of science and technology, oil and natural gas were made available at a cheaper rate and consequently interest in solar energy declined. In 1973 with the oil embargo and subsequent rise in oil prices, solar energy was identified as the best alternative to oil and natural gas.

The earth receives about one-half of the one billionth of total solar energy output. The solar constant varies between 1,398 W/m^2 and 1308 W/m^2, the mean value being 1,352 W/m^2. Achieving a solar future requires hard struggle against technical difficulties and vested interests. The most challenging complex and urgent need is to bring the solar energy cost and efficiency to a point, where it is an economical energy alternative for widespread use, and to foster creation of viable, commercial independent solar energy industry. Some of the solar energy technologies in use are solar concentrators and solar heaters, solar homes, solar pond, heliostats and air conditioners, the heat pump, solar chimney and solar cells.

Wind Energy

The driving force of wind is the unequal heating of the earth and the atmosphere. Hence, it is the second hand solar energy. Further, the characteristic flow patterns of wind are due to the earth's rotation. Despite the intermittent nature of wind pattern at any particular site, it is almost a constant year by year. In the middle age, traditional windmills were used extensively to mill grain, lift water for drinking and irrigation and voyages. However, the present interest in wind power is to generate electricity and heat. Due to the intermittent and variable nature of wind source, some form of back-up is necessary for supply in the interrupted period.

Micro-Hydel Projects

In conventional hydel projects, the construction of large dams alters both the downstream and lake area ecology. The submerged area along with its agricultural products, flora and fauna gets affected. The population of the area has to be removed and resettled, which causes disturbance and hardship. The completion of such large projects needs quite a long time. Hence, small hydro projects called mini-hydel or micro-hydel, which can be built on small downstreams and even on canals, are viable alternatives.

Tidal Wave

The tidal power is free, clean and inexhaustible. The tides cannot be exhausted till the moon—their source of energy exists. Large amounts of electric power could be developed in the coastal regions having tides of sufficient range, although, even if fully developed, this would amount to only a small percentage of the world's potential power.

Ocean Wave and Ocean Currents

The rhythmic up and down motion of the waves caused by the wind represents a huge potential source of energy. The trouble with wave power is that it is diffused thinly over a vast area.

The ocean currents—massive bodies of water moving endlessly along the same path due to prevailing wind and Coriolis forces—are another source of power that can be tapped easily. Again, the trouble is that the movement is relatively slow, about 8 km per hour.

Geothermal Energy

The heat of the earth in deep rocks, i.e., the geothermal energy can give rise to hot springs when water naturally accumulates underground. Springs are ephermal, discharging intermittently or permanent discharging constantly. A borehole drilled down into the water can be used to get a supply of hot water for industrial or domestic heating. Such systems have been already operational in Scandinavia, USSR and USA.

An artificial hot spring can also be used as a source of heat energy. A borehole drilled several hundred metres into a natural cavity in earth in which the temperature may rise to 300°C. The hot water pumped down the bore is led up a second borehole and at the surface the hot water is passed through a heat exchanger, which transfers its heat to the air blown over it.

In regions such as Iceland, USA and some places in New Zealand, which have active volcanoes, geysers spout steam and superheated water out of the ground. The energy of this hot water can be easily tapped.

Thermal Energy of Ocean Water

The temperature difference between the surface water and the water lying on the seabed below is anything up to 24°C. This is quite important due to the large volume of water involved. The ocean thermal energy conversion (OTEC) is not a futuristic idea, it works today. OTEC systems need a temperature difference of at least 20°C in order to work properly. Almost all the waters between the Tropic of Cancer and Capricorn could support an OTEC plant.

Alternative Resources

Magneto-Hydrodynamic (MHD) Power Generation

The MHD generator, which directly converts thermal energy into electricity, is a highly efficient heat engine with a high efficiency of 60 per cent compared to 35 per cent efficiency of conventional thermal power stations. The principle of MHD generator is based on Faraday's law of electromagnetic induction. In MHD generator, when an ionized gas flows across the lines of magnetic field, a voltage is induced. The gas used may have a temperature between 2,000 to 3,000°K.

It requires a working fluid, which is a partially ionized conducting gas. Gases become conductive when their atoms or molecules are stripped of one or more electrons thermally, electrically or by radiation. Hence, the conductivity of a gas depends on its temperature. However, to achieve thermal ionization of fossil fuels such as coal or inert gases, extremely high temperatures are required. Reasonable ionization and hence electrical conductivity is obtained around 2,000 to 3,000° K, when gases are seeded with additives of easily ionizing materials of alkali metals or potassium carbonate. Environmental pollution is considerably reduced in the MHD power generation.

Hydrogen

A number of scientists argue that instead of the wasteful burning of huge collection of fossils, it will be more practical to burn hydrogen. The amount of hydrogen stored in oceans is plenty. It is a component of coal gas and natural gas that is methane. Hydrogen was occasionally used to drive vehicles during World War II and presently it is already in use as a rocket fuel.

Energy Forest

The source of energy to be provided to the majority of people should be easily available, cheap and renewable. From this point of view, plantation of quick growing varieties of trees like Casurina and Euclyptus in the wasteland around the village in all Third World countries seems to be a practicable solution. It is estimated that the plantation of trees with 8-year cycle in 7.4 hectares of land can meet the energy needs of about 500 persons at the level of per capita consumption of 1 kilowatt of energy per day.

Biogas

In biogas plants, biological and cellulosic matter is fermented anaerobically with an arrangement to collect the generated gas, which is largely methane. The digested sludge with 1.5 to 2.3 per cent nitrogen is a better fertilizer than can be obtained through composting. The methane gas is a clean burning fuel and environment friendly.

Pyrolysis and Lucineration from Wastes

Pyrolysis is a process in which organic material is heated without air to temperatures of about 500°C, and as a result, a kind of crude oil is produced which has about 75 per cent of the calorific value of ordinary oil. The process can be economically justified when the energy costs are high, although quite a considerable amount of oil, that is, a barrel of oil per ton of organic wastes, could be produced in this way.

Energy Resources and Environment

The use of energy has definite damaging effects on the environment. We, for convenience, arbitrarily identify three phases, in which such damages can occur: during the exploitation for a resource, during production and processing of it and during consumption.

Impact of Exploration

The environmental impact of certain exploration depends on the location of energy reserves, the nature of the exploration process and the sensitivity of the affected ecosystem to perturbation. The energy resources, which are buried within the earth or under the seas, require substantial exploration. The present technology of geological engineering allows obtaining a great deal of information from existing maps and carefully placed boreholes. Hence, it is expected that the most serious environmental consequences of energy explorations is to be found only where local ecosystems are especially sensitive to the disturbance. The best example is the tundra in the Arctic region of Alaska, which has been disturbed and the disturbances will persist for decades due to the exploration of petroleum. In spite of small exploration bore-holes, the size of the equipment, the number of personnel involved in the exploration can cause considerable land disturbance with largely unknown consequences.

Impact of Production and Processing

In contrast to the relatively minor environmental impact of exploration, economically efficient production and processing techniques have large potential effects. For many kinds of energy such as hydroelectric, solar, tidal and wind, which have no substantial impact from exploration, the effects from production is either unknown or aesthetic in nature. In case of geothermal energy, it is speculated that the waste heat might escape to the environment and produce ecological damage. There are also possibilities of earth subsidence and enhanced earthquake hazard. In short, the potential environmental impacts of these forms of energy are largely unknown. All of them are cleaner than the fossil fuels and could contribute to improved environmental quality. However, due to the limitations in the possible energy output from these sources, they could provide only small amount of our requirement. Therefore, unless a breakthrough is achieved in direct utilization of solar energy, we must depend upon technology to clean up fossil fuels and confine the nuclear reactions, which are so important, yet potentially so damaging.

The fossil fuels often lie deep within earth's crust or under the ocean floor. To get the highest quality low sulfur anthracite coal, deep mine shafts must be excavated, which are costly, dangerous and produce vast quantities of waste rock. The remaining stocks of coal are mostly shallow deposits and hence strip mining is adequate. Strip mining requires no excavation, the overlying rock is removed entirely and the exposed coal then may be broken up and removed. The stripping is cheaper and is likely to be used even more widely when low quality coal deposits can be used more extensively as with MHD power generation. However, stripping is ugly and chews up landscapes, leaving big pits and mountains of overburden. It also disrupts the water tables. The environmental effects of oil production are particularly hazardous. In 1969, Santa Barbara oil spill and a number of marine oil tanker disasters have made the public aware of the dangers involved in the production and transportation of oil in aquatic environments.

In case of nuclear power, the environmental dangers are potentially the most hazardous. Fission reactions are accompanied by the escape of tritium into the environment. The radioactive lifetime of tritium is short but it is readily incorporated into biological systems. Thus theoretically, tritium can pose a genetic danger of unknown magnitude, although its emissions are not strong enough to produce direct radiation damage. In fact, tritium is used widely in research as a marker in living cells on the assumption that it is rather harmless and there is no evidence on the contrary. Other isotopes involved in fission have definite effects on the environment. Basic fuels and fission products are much more dangerous radioactive substances both because of their long half-lives and strength of emission. Moreover, the accidental explosions in nuclear power plant could spread radioactivity over large areas. The other set of environmental problem arise from the low efficiency of conventional nuclear power plants due to which a large amount of heat energy is wasted and must be disposed somewhere in the environment. The nuclear power plants cannot consume the entire amount of fuel completely and even when the critical minimum density of ^{235}U in a fuel rod is too low to sustain fission, the remaining material is still dangerously radioactive. The fuel rods must be reprocessed and waste product must be disposed of. Transportation of highly radioactive materials and disposal are quite problematic.

Some critics compare the fission reactors to time bombs and oppose their construction entirely. The nuclear scientists claim that the nuclear power plants are less dangerous than sum total of day-to-day accidents and such misconceptions are due to the lack of proper information.

There should not be any hesitation in the exploitation of nuclear energy because of its immense potentiality. However, there is no satisfactory conclusive agreement on this issue.

Impact of Consumption

In a general sense, the environmental quality is a measure of human beings and their work. The environment has been exploited for resources and reshaped toward our ends. We now discuss in the following various forms of environmental pollution: water pollution, air pollution, solid waste, thermal pollution, biocides and radioactive wastes.

The variety of pollutants that enter natural waters worldwide is vast. However, many of them are industrial and technological wastes. Further, domestic sewage and garbage routinely enter natural waters. In developed countries, the treatment plants and septic tanks are imperfect. In developing countries, untreated waters routinely enter streams. In oceans, oil slicks from ocean shipping, mark the major commercial sea-lanes and sewer pipes belch directly from coastal cities into the estuary and sea. Long-life biologically active pesticides run off the land into streams and rivers and finally into the sea. Water is an excellent solvent for all sorts of substances. Salts, dyes and acids can be carried far from the actual site of disposal. A striking example is the discovery of DDT in penguins in Antarctica, where it has been never used.

The most likely places, where air pollution is observed, are major cities, urbanized regions and major highways. In populated areas, locally high concentration of air pollution is a logical result of industrial emission, internal combustion engines and production economies. In rural areas, extensive agricultural burning creates dense smoke clouds.

The solid wastes include domestic wastes like bottles, cans, toothpaste tubes and automobile bodies as well as industrial wastes. In addition to these municipal wastes, one has mining wastes and wastes from agricultural sources. The radioactive waste management is another difficult problem.

The waste heat is responsible for thermal pollution. Automobiles, industries and nuclear reactors are major source of thermal pollution. The biocides are the pressing need to defend our beasts and our crops against the ravages of nature. However, the use of such biocides has resulted in detrimental effects on human population.

Ethical Problems

Now we are in a position to critically analyze the ethical problems of environmental pollution.

Pollution as both Natural and Cultural

A pollutant, seen in the perspective of our initial discussion in the chapter, is not something apart from humans but is inherent in their very biology and culture; it is a result of their peculiar adaptations and attributes. Pollutants are natural by-products of human manipulation of nature. The so called pollutants, a value and emotion laden concept in today's society, are in fact, normal by-products of people as purely biological organisms and as creative social beings. They are the organic and inorganic wastes of metabolic and digestive processes and of creativity in protecting and augmenting the production of crops, of warming homes, clothing the body and harnessing the atom. The problem is not the natural elaboration of by-products; it is in the disposition of them; also of a resource being out of place—too much of a resource in one system, not enough in another. The problem is further aggravated with a resource being present in a system that is not adapted to it and thus constituting an unaccustomed stimulus, stress or "insult" to the system. These are stimulants that may terminate some or initiate other biological processes, alter efficiency, affect species composition and structure and in general thereby alter the dynamics and development of an ecosystem.

By-products will always be with us. They will increase as technology and living standard increases, they will become exacerbated as urbanization proceeds and more people live in small areas. Solutions do not and cannot lie solely in removing the cause because as long as humanity exists, it will have by-products. Rather, the answer lies in intelligent management of that production through regulating the unfavourable alternation of our surroundings. Given that there are ultimate limits to the amount of these resources, it becomes an expression of wisdom to plan ahead to a steady-state system, not to an ever-expanding one. But we have moved too fast towards a solution without first having discussed specifically the effects of misplaced resources or pollutants, in context of actual ecological system.

Internal versus External Environment

The intuitive meaning of environment implies the physical as well as biological world surrounding us. We tend to be concerned only with the biosphere while considering the issue of environmental impact on us. Such a concern naturally focuses upon the spatial dimension of environmental impact on biological systems like us. What is overlooked in this way of viewing the environmental situation is the very important fact of the essentially periodic character of the way biological systems regulate their internal environment, as well as their relationship to the outside world. Thus, the biosphere together with the internal environment constitutes the total spatio-temporal dimension of the environment. The regulatory mechanism involves feedback cycle that generates oscillations. The natural selection may dampen or enhance the amplitude of these oscillations, but the periods have been adjusted to match closely the periods of astronomical events that are of importance to living beings. Both the internal and external features of the environment have provided nature with abiotic selection pressures that have shaped the evolution of life on earth. The adaptive significance of the internal environment is most likely to lie in the opportunities that it provides for the regulation of the phases of its components to each other as well as to the external environment.

Thus, an organism can be thought of as an ensemble of oscillations organized by complex internal programmes. With this perspective in mind, it is clear that tampering of the external environment results in changing the internal environment of the organism. However, adaptation is taken care of by the spatial change of the internal environment while the temporal changes in the internal environment result in catastrophic consequences leading to extinction of species. This clearly sets the limit to the extent external environment should be tampered with by the human activities.

Conclusion

We have analyzed the origin, use and availability of conventional as well as non-conventional energy resources. It is the law of nature, that the exploitation of energy resources for useful work results in thermal pollution. Other forms of pollution are also to large extent inherent in human biology and culture. Finally, we have shown that the existence of species is governed by the subtle play of spatio-temporal features of internal as well as external environment. Hence, the exploitation of environment to the extent that it affects the temporal dimension of internal environment is unethical.

References

E.J. (1986), *Concepts of Ecology*, Kormondy, PHI, New Delhi.

Foin, Jr., T.C. (1976), *Ecological Systems and the Environment*, Houghton and Miffin Co., U.S.A.

Teller, E. (1979), *Energy from Heaven and Earth*, W.H. Freeman and Co., San Francisco.

21

Role of Environmental Management Standards in Improving the Environment

Vishnu Ratna

Human activities in recent times have influenced the environment in a way that has endangered life on earth. Newton's third law of motion "to every action there is an equal and opposite reaction" fits appropriately in the context of the present scenario of environmental degradation. The exploitation of natural resources to fulfil human desires has ultimately resulted in nature's taking revenge in the form of several global environmental crises. The depleting resources, coupled with environmental imbalances, are now becoming a serious cause of concern for the survival of this planet.

Because of our inability to foresee the future, we have shaped our society so as to provide for the needs and desires of the current generation without any consideration of the implications that it could have for the future generations. The concept of sustainable development is an attempt to visualize impact that our present-day activity could have on future generations. The concept is very laudable; however, to bring it into practice is something that practitioners, professionals, governments and other responsible stakeholders have to think about quite seriously.

A holistic view of life considers the world as an integrated whole. The ecological view is that all phenomena are embedded in the cyclical process of nature. The ecological view is somewhat different from the holistic view in as much as it takes into account the impact of human action on the natural and social environment. Shallow ecology is anthropocentric. It considers man to be at the centre of the universe; everything else is secondary. Deep ecology encompasses every living being as a part of the natural environment.

Life is not isolated; it is interconnected. Edward Lorenz, a meteorologist, discovered "butterfly effect" that asserted that a small perturbation in a remote part of the universe, sometime in remote past, could be the cause of typhoon on earth today. Thus the new worldview considers a network of living beings rather than individual beings living in isolation. This consideration alone can help us to decide whether an activity is sustainable or not.

Environmental concerns are multifarious. Global warming, ozone depletion, depletion of water resources, bio-diversity, deforestation, and desertification are to name a few. But all the environmental issues are interconnected with each other. Imbalance in wealth in different parts of the globe could impact upon human activity to cause environmental degradation all over the globe. Lack of education and undernourishment for a vast number of human beings would cause population to grow, thereby affecting natural resources and land use to disproportionate levels.

'Gaia' Theory

Dr. James Lovelock, a British chemist specializing in the atmospheric sciences, worked on a NASA programme for the evidence of extra-terrestrial life on Mars. The analysis for Mars yielded this result: carbon dioxide (95.3%), nitrogen (2.7%), argon (1.6%) and only (0.15%) oxygen with (0.03%) water. In comparison, the earth's atmosphere at present is 77 per cent nitrogen, 21 per cent oxygen with traces of carbon dioxide, methane and argon.

"What was happening upon the earth which enabled the maintenance of such an unlikely combination of chemical gases—specifically nitrogen and oxygen? What complex processes are at work within the terrestrial atmosphere—and have occurred for many billions of years—to explain this uniqueness? How have these processes arisen and what today maintains these processes at this equilibrium which is chemically *far from equilibrium*?" he questioned.

Lovelock took the first steps in answering these questions by considering the beginnings of life upon the planet earth. The earliest of life forms existed in the ancient oceans and were the smallest and the simplest—less than single celled. Contemporary microbiological research points to the fact that almost 3 billion years ago, bacteria and photosynthetic algae began extracting carbon dioxide from the atmosphere and releasing oxygen back into it. Gradually—over vast geological time spans—the atmospheric chemical content was altered away from the dominance of carbon dioxide towards the dominance of a mixture of nitrogen and oxygen—towards an atmosphere, which would favourably support organic life powered by aerobic combustion—such as animals and mankind.

The resultant term 'Gaia'—after the Greek goddess who drew forth the living world from Chaos—was chosen.

By 1979, James Lovelock had published some of his ideas in a first book *Gaia: A New Look at Life on Earth* in which the statement of the specification of the Gaia hypothesis had become somewhat better defined. Lovelock postulated that the entire range of living matter on earth from whales to viruses and from oaks to algae could be regarded as constituting a single living entity capable of maintaining the earth's atmosphere to suit its overall needs and endowed with faculties and powers far beyond those of its constituent parts. Gaia can be defined as a complex entity involving the earth's biosphere, atmosphere, oceans, and soil; the totality constituting a feedback of cybernetic systems which seeks an optimal physical and chemical environment for life on this planet.

Standards and Regulations

To bring back normalcy from the imbalance created by the human activity, concerted efforts are required by the communities in various countries. One method is to use regulations and enforce strict measures for compliance. Governments make regulations and the industry associations, consumer associations and other stakeholders make voluntary standards. Failure to meet the statutory regulations attracts penal provisions under the law of the land. No such penalties are, however, imposed in not fulfilling the needs of the voluntary standards, but their adoption does provide the benefits for harmonious trade and commerce. For example, a regulation may specify the maximum content of sulphur in the smoke emission. However, it is left to industrial standard to define how the sulphur content is to be measured. Formulation of voluntary standards is an aid to the regulatory authorities in limiting regulations and it also helps the enterprises avoiding proliferation of statutory regulations. It is for this reason that standards have to evolve from time to time to cater to the changing needs of the society.

Evolution of Management Standards for Quality and Environment

ISO 9000 Quality Management Systems

International Organization for Standardization (ISO) is a non-governmental body, and its standards are purely voluntary. However, because ISO standards are developed in response to market needs, and are based on consensus among the parties with a direct interest in their contents, the incentive for implementation of ISO standards is high.

In 1987, ISO released 9000 series of quality management standards, which were identical to the BS 5750 standards, adopted in UK since its release in 1979. ISO 9000 is one of the most successful series of standards in the history of the ISO. Thousands are implementing it for manufacturing and service organisations both in public and private sectors. The ISO 9000 series of standards were revised in 1994 and the latest revisions were released in 2000. More and more suppliers worldwide are using the standards to judge the quality of their suppliers. As the trend builds, certification will grow increasingly more attractive for manufacturers that want to conduct trade in the global marketplace.

Standards for Environmental Systems

Evolutionary trends in the development of environmental systems have been closely related to the trends in the area of quality management systems. With the growing concern internationally towards environmental issues, each country has formulated its own regulatory norms to control the environment. However, regular monitoring of these norms by the environmental regulatory bodies is akin to the age-old inspection system of the quality discipline where the purchaser inspected the product after its manufacture. Just as it was found by the quality community that mere inspection of the product subsequent to its manufacture did not give assurance that all that was required had been addressed to achieve the required level of quality, similarly in controlling the environment, full assurance is not possible unless a system is designed around sound management principles to which an organization demonstrates commitment on a continuing basis.

An Environmental Management System (EMS) may be defined as "that part of the overall management system which includes organizational structure, planning activities, responsibilities, practices, procedures, processes, and resources for developing, implementing, achieving, reviewing, and maintaining the environmental policy" of an organization. An effective EMS can help an organization manage, measure and improve the environmental aspects of its operations. An EMS recognizes that almost all activities performed by an organization may have an impact on the environment.

BS 7750 *Environmental Systems*

Close on the heels of quality management systems, came BS 7750 Environmental Management System (EMS). This standard was introduced in UK in 1992 and revised subsequently in 1994. This standard is also generic and is designed for all types and sizes of organizations—from a chemical processing plant to a school, from an engineering company to a local government department. Like ISO 9000, it is a voluntary standard, which sets out systems to protect the environment through a quality management approach. BS 7750 specifies requirements for the development, implementation and maintenance of an EMS. It does not include certification arrangements, but does provide for the requirements against which certification can be performed.

BS 7750 is system-oriented. It is based on 11 main clauses and 11 annexures, which set out what environmental management, should be. It is a document-based system with supporting procedures designed to achieve continuing improvements in environmental performance. It provides a means for enabling an organization to reduce and control the impact of its activities on the environment. As with quality and, health and safety, the most effective path to improvement is to incorporate control into the management of the organization, making it part of the day-to-day decision-making process. The guiding principle in the EMS is prevention of error and a proactive approach on the part of the organization in identifying the areas of concern and effective handling of the environmental issues.

Eco-Management and Audit Scheme (EMAS)

This is a regulation within the European Union (EU) agreed on by all member-states. EMAS became formally operational in May 1994 and is based on voluntary publication of verified environmental performance data. Participants will make a public statement of their current performance and future aims, by defining specific values achieved and objectives, which they intend to meet, and an independent body will verify that public statement. To comply with the regulation, the organization must also operate an environmental management system, which can be demonstrated to control the programme under which those specified objectives are managed. In Europe, companies are now gearing up to compete on the basis of environmental performance improvement. The frequency of environmental audits to be undertaken for specific sites is prescribed in the regulation and ranges from one year for a site with high environmental impact to three years for a site with low environmental impact. There are two essential differences between BS 7750 and EMAS:

First, BS 7750 is a UK industry specification, which aims to establish an EMS for any business sector. EMAS, on the other hand, is a European-wide accepted scheme only open to organizations with an industrial activity, and to local authorities. Second, EMAS requires an environmental statement designed for the public and written in a concise, and comprehensible form.

ISO 14000 Environmental Systems

The world is today moving towards a common market for which efforts are on in developing harmonized standards, which are acceptable to all the nations for conducting trade and commerce globally. This task is accomplished by ISO whose membership today comprises the principal national standards of some 116 countries. This international body, which has completed its 50th anniversary in 1997, has closely been working in evolving a consensus standard for environmental management. The requirements of BS 7750, EMAS and of various regional trade blocks of South East Asia, Europe and Latin America are, therefore, being addressed in the ISO standards for adoption worldwide. ISO 14001 was developed around the principles of BS 7750, although the order of clauses is different and the standard is more open to interpretation. ISO 14001 is the specification standard and, consequently, provides the basis for a third party certification. Like ISO 9001/9002/9003 standards, ISO 14001 could be mandatory as part of a two-party agreement (e.g., contract). ISO 14004 provides guidance for an optimal management system.

Initially, 18 standards have been proposed by ISO to encompass the following: (a) Environmental Management Systems (EMS), (b) Environmental Auditing (EA), (c) Environmental Labelling (EL), (d) Environmental Performance Evaluation (EPE), and (e) Life Cycle Assessment (LCA)

In respect of EMS and EA, the ISO released the following five standards in 1996:

- ISO 14001: 1996 "Environmental management systems—specifications with guidance for use".

- ISO 14004: 1996 "Environmental management systems—general guidelines on principles, systems and supporting techniques".
- ISO 14010: 1996. "Guidelines for environmental auditing—general principles".
- ISO 14011: 1996. "Guidelines for environmental auditing—audit procedures—Auditing of environmental management systems".
- ISO 14012: 1996. "Guidelines for environmental auditing—qualification criteria for environmental auditors".

ISO 14001: EMS Specification Standard

The key requirements contained in ISO 14001 in regard to EMS standard are: (a) definition of organization's environmental policy; (b) planning, including consideration of: environmental aspects, legal and other requirements, objectives and targets, and environmental management programs; (c) implementation and operation, including structure and responsibility, environmental management system documentation, and emergency preparedness and response; (d) checking and corrective action, including monitoring and measurement, and environmental management system audit; and (e) management review.

ISO 14004: EMS Guidelines Standard

ISO 14004 includes a model for an EMS which subscribes to the following principles: (a) commitment and policy, (b) planning, (c) implementation, (d) measurement and evaluation, and (e) review and improvement.

ISO 14010: General Principles for Environmental Auditing

These general principles for environmental auditing have been enunciated in ISO 14010 and include: (a) clearly defined scope and objective; (b) objectivity, independence and competence of auditors; (c) the use of audit criteria, collection of audit evidence, and documented findings; and (d) reporting the audit findings in a written report.

ISO 14011: Audit Procedures for EMS Audits Leopard

ISO 14011 provides procedures for conducting EMS audits which are defined as "systematic, documented verification process of objectively obtaining and evaluating audit evidence to determine whether an organization's EMS conforms with the EMS audit criteria, and communicating the results of this process to the client". The following audit elements have been specifically focused on EMS: (a) defining roles, responsibilities and activities including lead auditors, auditors, the audit team, the client and the auditee; (b) performing the audit, including initiating the audit and defining its scope, preparing the audit plan and making audit team assignments, conducting the actual audit (opening meeting, collecting evidence, etc); and (c) reporting the results of the audit, including defining the content of the audit report and preparing and distributing the audit report.

ISO 14012: Auditor's Qualification Criteria

ISO 14012 was developed to provide guidance on qualifications appropriate for the lead auditors and auditors. The standard provides guidance on the qualification in the following areas: (a) education and work experience, (b) auditor training, (c) personal attributes and skills, and (d) language and communication skills. The standard also includes guidance on the development of an environmental registration body. It may also be applied to internal auditors for use by the small and medium enterprises, which may not be able to afford external auditors in every situation.

Environmental Performance Evaluation (EPA)

Evaluation of environmental performance is defined in draft ISO 14031 standard as "review of an organization's environmental aspects to determine whether objectives are met". This includes any activity of the organization that can interact with the environment including emissions, effluents, and disposable waste or consumption of energy, water, land, and other natural resources.

Environmental Labelling (EL)

ISO labelling programme has arisen on account of non-standard use of methods used in many countries, which is seen as a trade barrier for international trade. Five standards are proposed in this area; ISO 14020 is a general principle standard, which apart from other things specifies that the labels should be accurate, verifiable, relevant, and non-descriptive. It should not create unfair trade restrictions in domestic and foreign products and services. In addition to this, certain specific claims of manufacturers, e.g., recyclable, biodegradable, etc., are defined in the standard. Use of symbols for labelling is also under development.

Life Cycle Assessment (LCA)

LCA is a methodology for analyzing a product or service through all stages of its life cycle: raw material acquisition, manufacturing, transportation, use/reuse/maintenance, and recycle/waste management/

disposal. The ISO 14040 and ISO 14041 standards include LCA methodology with such elements as life cycle inventory analysis, and life cycle impact assessment.

Socio-Economic Benefits of EMS Audit

Free trade is now a well-established and widely advocated economic philosophy. Numerous free trade areas are emerging and rules governing trade across the globe are under formulation. The World Trade Organization (WTO) is clearing numerous hurdles imposed individually by the nations in conducting free trade. For free trade to succeed, it must be implemented with a view to improving the earning power, living standards, working conditions, quality of goods and services, quality of life, and quality of environment for most of the population in all the countries involved. The countries can reap the socio-economic benefits of a free trade only through harmonization of standards across the globe. In the past, trade restrictions were created through formulation of policies, which suited the interest of a particular nation most, but now sovereignty is gradually being sacrificed through common goals. Issues of quality, environment, safety and health are no longer matters for discussion and debate within the boundaries of a particular country it now concerns everyone on the globe.

EMS requirements go beyond the goals of just meeting the regulatory requirements; they seek to establish environmental policy and objectives, achieve compliance with them, assess the effectiveness of the EMS and check the continuing relevance of the environmental policy. The need for maintaining the ecological balance through environment-friendly production is being widely recognized as important. ISO 9000 has succeeded primarily because customers realize that quality is essential. Similarly, with the growing concern for environment, the customers are regarding EMS as an important consideration for free trade.

The driving theme of the EMS is evaluation of the activities of the organisation influencing the environment in a significant way, and then manages their effects by setting up objectives or applying controls. It calls for documenting a 'Register of Environmental Effects' for identification of environmentally significant issues.. Environmental effects can be categorized according to whether they pollute something into the environment or are of a resource depleting nature, i.e., taking something out of the environment. A polluting effect may relate to air, land or water, e.g., acid gases, particulates, pesticides, heavy metals, sound, and heat. A resource usage effect deprives the environment of the vital or the necessary resources, e.g., fossil fuels, water, wood, minerals and land. The EMS requirements also call for maintaining a register of legislative requirements since that is the baseline for any organization to launch a programme of improvement. However, if an EMS is purely legislation oriented it is most likely that full benefits may not be derived and business opportunities may be lost. A good EMS should, therefore, be based on managing the significant environmental effects. Further, an EMS does not regard controlling and improving the pollutants but also considers the way an organization handles resource and energy extraction for its operation. Environmental objectives and targets are therefore to be formulated in the same way as other business objectives. The effects on environmental matters by an organisation's supply chain as well as the distribution chain are also to be taken into consideration during implementation of an EMS programme. This necessitates evaluation by an organisation of its supplier's operations in a meaningful way and has a snowballing effect in building up environmental concern much beyond the scope of the organisation's own activities. Benefits of EMS programmes implemented in organisations in Europe have been reported. HB & Associates, a British consulting firm launched an EMS programme in "Beacon Press". Apart from marketing advantage, it resulted into cost savings from reduced consumption of waste, reduced insurance premia due to lower risk involved, avoidance of environmental mishaps, and cost savings from anticipating future legislation and regulations. Contrary to the increasing importance given to the EMS in the developed countries, the level of awareness in the developing countries is still much below the expectation. Parikh and Sharma (1996) have conducted study of the Indian tanneries from the point of view of meeting the changing international environmental regulations.

The industry is gradually losing business in the international market due to non-fulfilment of changing environmental criteria. The demands for use of eco-friendly chemicals in leather and textiles is increasingly being insisted upon by the importers. To provide assurance to the importing buyers it would, therefore, be more important now to build up a system based on the EMS criteria than on the basis of the product or process standard alone. The key feature of EMS is to implement a programme of regular audits. An audit is an independent review conducted to compare some aspect of environmental performance with a standard for that performance. An audit may cover both internal and external systems for achieving the level of performance. The role of an auditor is that of a professional providing constructive inputs, guidance and continued emphasis on the environmental matters within the organization. Without this constructive component, the environmental audit takes on the traditional role of audit only, which is frequently perceived as punitive or negative. If the environmental audit is used as an educational tool and enhancement to the total environmental management programme, the benefits are far-reaching and the programme can grow.

In the emerging scenario, meeting the product and process standards alone would not suffice to gain the competitive edge in the international trade; implementation of a formal EMS would be increasingly emphasized. The Indian industry would have to gradually prepare itself to take up a more proactive approach through implementation of EMS if it has to survive in the free trade economy of the world market.

References

Benett, Stephen, Environmental Management Systems, *Quality World*, May 1995, p. 325.

Fritjof Capra (1997), *The Web of Life*, Flamingo.

Harris, Terry, Environmental Reward for Beacon Press, *Quality World*, September 1996, p. 596.

Lovelock, James (1972), "Gaia as Seen Through the Atmosphere", *Atmospheric Environment*, Vol. 6, p. 579.

Parikh and Sharma (1996), "Economic and Policy Analysis of Trade and Environment Linkages in India", *Journal of IAEM*, Vol 23, pp. 71-78.

Struebing, Laura, 9000 Standards?, *Quality Progress*, January 1996, p. 23.

22

Degradation of Groundwater Quality Due to Arsenic, Nitrate and Fluoride Pollution

Jyotsna Lal

Drinking water is one of the basic necessities of life. It has been observed that water is a major source of pollutants and contaminants, which cause several ailments. Large number of people runs the risk of suffering the adverse effects when water is unsafe to drink. WHO estimates indicate that 80 per cent of the diseases are associated with contaminated water. The common diseases are hepatitis-A, polio, typhoid, cholera, dysentery, dental and skeletal fluorosis, methaemoglobinemia.

The presence of disease causing micro-organisms and/or excessive dissolved compounds and salts, metals like fluorides, nitrates, iron, arsenic lead, chromium, mercury, cadmium, copper in surface and groundwater sources leads to contamination of potable drinking water. Groundwater is severely threatened by pollution due to industrial wastes and excessive inputs of population. Large-scale industrial growth has caused serious concern regarding the susceptibility of ground-water due to heavy metals. Waste materials near the factories are subjected to reaction with percolating rainwater and reach the groundwater level. River water pollution is mainly through industrial wastewaters from paint, pigment, chrome and leather tanning, electroplating, textile dyeing and ceramic industries.

There are number of toxic substances found in natural and waste-waters, some of these are essential at low concentrations, serving as nutrients for animal and plant life but toxic at higher levels. The permissible limit for these toxic substances in potable waters is given in Table 1.

Table 1

Some Toxic Substances	Their Source	Permissible Limits (mg/l)
Arsenic	mining, pesticides chemical waste	0.05
Aluminium	alum added in water	0.05
Barium	treatment plants industrial wastes	1.00
Cadmium	metal plating, mining, industrial waste	0.01
Chromium	chrome plating and tannery effluents	0.05
Copper	metal plating, leaching, mining, industrial waste	0.01
Cyanide	metal plating, leaching, mining, industrial waste	0.01
Fluorine	natural sources, industrial wastes	1.50
Silver	photography	0.05
Lead	plumbing, mining, industrial wastes	0.05
Mercury	pesticides, coal mining, industrial waste	0.01
Manganese	natural sources, industrial wastes	0.05
Nitrate	nitrogenous fertilizers, manures, municipal and industrial wastes	45
Selenium	industrial wastes	0.01
Zinc	metal plating, industrial waste, plumbing	3.0

National Status of the Problem

The survey studies have revealed that the major problem being faced by our country is in the areas of arsenic, nitrate and fluoride contamination. Arsenic contamination has been found widespread in different regions of West Bengal due to dissolution of arsenic containing bedrock. The high arsenic content of the drinking water has become a major public health issue in 6 districts of West Bengal where an estimated 30 million people are dependent on tube-well water sources have been adversely affected. The districts are in the eastern sector of West Bengal, extending 450 km from north to south bordering Bangladesh. The severely affected districts are Malda, Murshidabad, Nadia, Bardhaman, 24 Parganas (North) and 24 Parganas (South). The arsenic content in the ground water ranges from 0.01 to 0.59 mg/l in as many as 830 villages in this area. Tube wells have been dug to a depth of 120 to 150 feet. The normal amount of As in hair is 0.08-0.25 mg/kg whereas 1mg/kg is an indication of toxicity. The normal amount of As in nails 0.043-1.08 mg/kg The normal amount of As in urine range 0.005-0.40 mg/kg [1.5 D] Maximum permissible limit of As for drinking water is 0.05 mg/l. Table 2 shows the arsenic levels in drinking water in West Bengal.

Table 2a
Occurrence of Arsenic in the Groundwater of West Bengal

8 districts/830 villages in 58 blocks	Conc of	As in water	mg/l
Total samples analysed	< 0.01	> 0.05	> 0.01
34900	43.5	38.7	56.5

Table 2b
Study Arsenic Level in the Human Tissue [mg/kg] and Body Fluid [ug/1.5D]
of the People of West Bengal

	Total samples analysed	% of samples above normal level
HAIR	3530	74.0
NAIL	3620	74.0
SKIN	163	[1.2-29.3]
URINE	7240	83.6

High levels of nitrate (permissible limit 45 mg/l as NO_3^-) are a serious problem in different regions of the country, mainly due to high application rate of animal manure and artificial fertilizers. Summarized information on the occurrence of nitrate in ground water in India is given in Table 3.

Table 3
Summarized Information on the Occurrence of Nitrate in the Groundwater

State	District	nitrate mg/l
Andhra Pradesh	Hyderabad	11.8.7
	Secunderabad	63.3
	Visakhapatnam	1.12-28.9
Assam	16 districts	0.35-3.2
Harayana	Mahendragarh	> 100

Contd...

Contd...

Karnataka	Bangalore	>50
Maharastra	Nagpur	>45
Orissa	Ganjam	3.6-800
Rajasthan	Barmer	>11.3
	Bikaner	..
	Jaipur	..
	Jaisalmer	..
	Udaipur	..
	Jodhpur	45-613
Tamil Naidu	Chennai	15-45
Uttar Pradesh	Lucknow	0.0-650
	Jaunpur	65-1250
	Varanasi	59-284

Fluorosis is a crippling disease affecting nearly 25 million people in the country. The permissible limit for fluoride in drinking water is 1.5mg/l Fluorosis problem is widespread all over the country. The endemic states are Andhra Pradesh, Karnataka, Tamil Nadu, Uttar Pradesh, Haryana, Punjab, Maharashtra, Gujarat, Rajasthan, Jammu & Kashmir, Delhi and Kerala. Table 4 and 5 give summarized information on the occurrence of fluoride in groundwater in the different states of India.

Table 4

Summarized Information on the Occurrence of Fluoride in Groundwater in Various States of India

State	District	Fluoride mg	litres
Andhra Pradesh	Ananthapur	0.8-3.5
	Nalgonda,Prakasam Visakhapatnam	0.50-7.5
	Hyderabad	0.3-1.9
Bihar	Palamu	2.24-7.54
	Jamui	2.96-8.16
	Gadhwa	1.60-4.8
	Bhabhna	1.80-3.0
	Rohtas	2.50-3.0
	West Champaran	1.60-2.0
	Gopalganj	1.60-2.5

Contd...

Contd...

Gujarat	Mehsana	1.58-9.9
Haryana	Gurgaon	0.17-24.5
Jammu & Kashmir	Doda	0.05-4.21
	Ghat, Monkhil, Malwas, Bhagwa, Khastigarh, Bharat, Chamalwas, Batroo	
Karnataka	Dharwad	0.40-18.0
	Gulbarga	0.20-5.60
	Raichur	0.40-8.5
Maharastra	Jalgoan	0.11-3.0
	Bhandara	1.50-10.2
	Nagpur	0-4.4
M.P.	Shivpuri & Jabua	1.50-4.2
Punjab	Sangrur	0.28-4.0
Tamil Nadu	Madurai	0-0.27
	Dindigul	0.2-1.03
	Tuticorin of Chidambarar	above 0.5
U.P.	Unnao	0.12-19.0
	Agra	0.28-22.0
	Allahabad	1.00-1.50
	Dehradun	0.02-0.10

Industries Releasing Arsenic, Nitrate and Fluoride

Arsenic is used in metallurgical industry, glassware and ceramic industries, dye and pesticide manufacturing industries and petroleum refineries. Chemicals containing Arsenic are used in manufacture of herbicides and pesticide In West Bengal, a factory manufactures acetocopper arsenite, a pesticide, polluted drinking water in the southern part of Kolkata. Arsenic is released into the surface water through mining and burning of coal as well as copper smelting. Rivers flowing through the coalfields of Bihar have been reported to carry large amounts of arsenic, responsible for downstream arsenic poisoning in West Bengal. According to studies, the arsenic contamination of the Damodar in West Bengal is mainly due to dumping wastes from the coalmines along the riverbed.

High levels of nitrate (permissible limit 45 mg/l as NO_3^-) are posing serious problems in different regions of the country; excessive use of nitrogenous fertilizers in agriculture has been the primary source of high levels of nitrate. The excessive concentration of nitrates in drinking water could be due to intensive animal operation, over application of animal wastes, soil induced mineralization of soil organic nitrogen, septic tank systems and land fills. Apart from nitrate, nitrogen is applied in ammonium NH_{4+} and amide NH_2. forms, which generates nitrates in soil system through mineralization which is fairly rapid in tropical and subtropical soils. Due to its high solubility in water and low retention by soil particles, nitrate is prone to leaching to the subsoil layers and ultimately to the ground water if not taken up by plants or denitrified to N_2O and N_2.

Industries using hydrofluoric acid or fluoride salts in their technology, i.e., aluminium industries, steel and enamel, pottery and glass industries, oil refineries, pharmaceuticals and cosmetic industries which liberate flourous gases to the atmosphere and are the main cause of industrial induced fluorosis. Fluoride in lithosphere and soil are found chemically in compound forms, i.e., flouraphite, fluorite and other rock forming minerals in order of (0.06%-0.09%) of the earth crust. Salt deposits of marine origin also contribute fluoride to the soil. The fluoride accumulation in soil is governed by: (i) natural solubility of fluoride compounds, (ii) acidity of the soil, (iii) presence of other minerals and chemical compounds, and (iv) amount of water present. The fluoride accumulation in soil is also dependent on depth factor.

Water Pollution Control

Availability of safe and potable drinking water still remains a distant dream for a significant percentage of rural population of India. Rajiv Gandhi National Drinking Water Mission was introduced by the Government of India in 1986 with the objective of improving the performance and cost-effectiveness of the ongoing programme on drinking water.

Portable low cost water quality testing equipment has been developed by Industrial Toxicology Research Centre, Lucknow, National Environmental Research Institute, Nagpur and All India Institute of Hygiene and Public Health, Kolkata for detection of nitrate, fluoride and arsenic in drinking water samples obtained from tube wells, hand pumps of rural India.

Water treatment technologies and purification methods are being tested and several pilot projects have been set up by Department of Science and Technology in rural Rajasthan. Nitrate can be removed effectively by deionisation desalination, reverse osmosis, biological denitrification and electrodialysis. A study indicated that activated alumina could achieve over 95 per cent arsenic (V) and (III) removal twice as effectively as activated bauxite and 12 times more effectively than activated carbon. Activated alumina removed As (V) over more effectively than As (III) at pH 5 activated alumina removed also Fluoride along with Arsenic. Inexpensive defluoridation units have been tested in laboratory and field application. Ion exchange, Nalgonda technology, and absorption on activated alumina for fluoride removal have been reviewed in this light. Water treated by Nalgonda technology contains residual aluminium in. the range of 2.1-6. 8-mg/l under various operating conditions. Al in concentration of 8mg/l or more results in Dialysis dementia and risk of Alzheimer's disease. Activated alumina has been extensively used both for laboratory and field application. Activated alumina is widely used as deflouridation material in developed countries. UNICEF sponsored the development of domestic and handpump attachable defluoridation units using activated alumina. The study area for field performance was Unnao district near Kanpur, UP.

Health Hazards Due to High Arsenic, Nitrate and Fluoride Intake

As (III) ion causes arsenicosis in human beings. Symptoms of arsenic poisoning are abdominal pains, vomiting, diarrhoea and pain in the extremities followed later by numbness and tingling of the extremities, palmoplatar hyper keratosis, mees lines on the finger nails, deterioration in motor and sensory responses. Chronic arsenicalism include dermal lesion, peripheral neuropathy, skin cancer, peripheral vascular disease. Bony fishes and fresh water mussels contain a high content of As.

As, Hg, Pb, Cd are highly toxic even at low concentrations. Cd (II) substitutes Zn (II) in some metallo-enzymes. Arsenic acts on the biological system in the following ways: Trivalent toxic As inhibits enzyme activity by reacting with ligands containing sulphydryl ligands. As (III) attacks SH groups of an enzyme, i.e, inactivation of pyruvic dehydrgenase by complexation with As (III) preventing the generating of ATP in citric acid cycle. Pyruvic dehydrgenase is highly sensitive to As because of its interaction with two SH groups of lipoic acid, leading to pyruvate accumulation. As (III) ion readily crosses the placental barrier and causes foetal damage. Arsenic when absorbed in the biological system can undergo bio-transformation under in vivo conditions. As (V) ion or arsenate can be reduced into As (III) ion or arsenite. The half-life of this bio-transformed arsenite is three to five days, when it can cause harm to the system.

Nitrate itself is not toxic. The presence of excessive quantity of nitrate in drinking water causes infantile methaemoglobinemia in babies under three months of age. Nitrate becomes a problem only when it is converted to nitrite in the human body and causes methemoglobinemia, alternately called 'Blue Baby Disease' or 'Blue Baby Syndrome'. Several studies have correlated nitrate in drinking water with gastric cancer. Nitrite produced from nitrate in drinking water enters the bloodstream mainly through the upper gastrointestinal tract. Almost all nitrate is taken but the efficiency of the process depends on the food matrix. Nitrate excretion in saliva and urine reaches a peak 3-6 hours after nitrate ingestion. Excretion mainly occurs through urine, sweat and faeces Bacteria in the mouth and gastric mucus transform part of NO_3^- to NO_2^-. Urine is mainly sterile but nitrate can form due to urinary infections.

For haemoglobin to act as an oxygen carrier, the iron atom within the molecule has to be in the reduced [FeII] state. When nitrite is absorbed in bloodstream, it oxidizes haemoglobin to methaemoglobin, a [FeIII] compound with reduced oxygen transport capacity, as this compound contains iron in its highest oxidation state which is incapable of binding oxygen. In this way, nitrate reduces the total oxygen carrying capacity of blood. As different parts of the body get deprived of oxygen, clinical symptoms of oxygen starvation start to appear; the skin starts to take a blue colouration wherefrom the disease derives its name, the Blue Baby Syndrome. In infants, where the diet is normally carbohydrates, coliform organisms are thought to be primarily responsible for conversion of NO_3^- to NO_2^- in the digestive tract. In adults, as the stomach fluid is more acidic, the nitrate reducing bacteria live in the lower intestine, therefore absorption of nitrite to the blood stream does not occur. Gastrointestinal illness and diarrhoea may allow the bacteria responsible for conversion of NO_3^- to NO_2^- to migrate from lower intestine to upper intestine and stomach, and increase the chances of nitrite formation prior to absorption in the small intestine.

Dental fluorosis occurs when the level of fluoride in water exceeds 1.5 mg/l. Teeth due to their high calcium content easily take up fluoride. Prolonged intake of water having fluoride levels exceeding 3 mg/l leads to skeletal fluorosis which affects children as well as adults. Symptoms of this disease are tingling sensation in the legs followed by pain, stiffness of the back.

Skeletal fluorosis leads to bow legs and knock knees In patients showing effects of skeletal fluorosis, the medical reports suggest accumulation of fluoride ion more in cancellous bones than in cortical bones, because of porosity, copious blood supply, presence of trabecular bone surfaces and bone engulfing bone marrow The fluorosed bone shows characteristic structural changes, viz., increased bone mass and density; exostosis(bone outgrowth) at bone surfaces; increased osteoid seam and resorption surfaces; increased trabecular bone volume, cortical porosity and periosteocytic lacunar surface; formation of unmineralized cartilaginous loci within the trabeculae of the cacellous bone; maximum ill effects are detected in the neck spine knee pelvic and shoulder joints and small joints of hands and feet; severity of skeletal flurosis, increases pain which is associated with rigidity and restricted movement of cervical and lumbar spine, knee, pelvic and shoulder joints; further severity results in stiff spine, i.e., bamboo spine, immobile knee, pelvic and shoulder joints; crippling deformity further results: kyphosis, scoliosis, hexion deformity, paraplegia, quadriplegia and osteophytes. Certain beverages like dry tea leaves contain 39.8-68.6 ppm fluoride. Chewing items like supari tobacco, pan contain 3.8-38.0 ppm fluoride. Use of fluoride bearing toothpaste and mouthwash lead to fluorosis. The accumulation of fluoride in the human body is due to the high reactivity of fluoride ion with calcium of teeth bones resulting to farm calcium flourphosphate (florapatite) crystals and leaving unbound calcium in certain locations in the same tissue, which gets calcified and in turn results in stiffness of tissues and joints causing skeletal fluorosis in late stages. The chemical substitution of fluoride ions replacing hydroxl ions and calcium hydroxy—appetite in bones is due to strong affinity between fluoride and biological appetite of the body once the appetite/fluorhydrxyapatite forms, it remains chemically stable until the tissue is reabsorbed or metabolized a little of fluoride increase in the body by diffusion and absorbtion. Nearly 90 per cent of fluoride in the body is associated with calcified tissues. The fluoride effects on proteins and on DNA molecules are also possible.

Conclusion

Water pollution has emerged as one of the most alarming public health problems in India in recent years. Based upon the above literature survey the following conclusion can be drawn as factors responsible for spread and severity of arsenicosis, methaemoglobinemia and fluorosis:

(i) Population overgrowth, consumption of more water for daily use; unplanned and indiscrimination in digging of wells; increase in rate of industrialization; high level of As, F-intake through drinking water and food stuff.

(ii) Methaemoglobinemia is mainly due to high application rate of animal manure and artificial fertilizers.

(iii) Several technologies have been studied and the process of absorption on alumina has an edge over other methods due to its simplicity and sludge free operation. Activated alumina has been widely recommended as a water treatment material in developed countries.

(iv) Alum based defluoridation technologies should be avoided. It has been observed that, there is a pressing need for development and introduction of simple and inexpensive methods of water treatment.

(v) To deal with the problem, awareness should be created among villagers about the ill effects of drinking contaminated water. There should be transfer of water treatment techniques to the villages. There is a need for water quality surveillance. If villagers are trained to use water-testing kits to monitor the purity of water independently, they will realise the gravity of the problem. The water pollution menace can be overcome by community participation.

23

Dilemmas of Sustainable Development in India

A.K. Sharma and P. Vigneswara Ilawarsan

Sustainable development is a major concern of all states and civil societies. It has now become more vivid than ever before that the levels of development are difficult to maintain. Unless well-thoughtout interventions are made in the development processes, a reversible process of de-development may be set in motion and development may not be sustained. However, there is no consensus regarding the approach to sustainable development. As a matter of fact the concept of development has itself become quite problematic. The reason is that it has both normative and positive elements. While some economists may still define it in terms of growth rate of per capita income, now there is a move to define it more in terms of social justice, freedom, or improvement in factors such as education, health, employment, and the condition of the disadvantaged sections. In this situation it may be necessary to examine the concept of sustainable development, the opportunities it offers, and the threats it faces in the wake of the recent developments in economy and society.

This paper defines sustainable development as society's capacity to improve or maintain the current levels of development. It aims at identifying certain dilemma of development. It goes beyond the environment-based concept of development, which has not been acceptable on the ground that it denies opportunities to the vast section of the underdeveloped. The future of development will depend on what choices are made and what alternatives are selected.

What is Sustainable Development?

In the mainstream discourse on development, to develop is to control natural and social environment for achieving continuously higher levels of material standards. Development perspective is a homocentric perspective. According to this perspective, the gains of development are eventually diffused to all. The expectation is that development would lead to improvement in health, better employment opportunities and social development. In 1950s, social and economic theories talked mainly about how to develop. The countries of Asia, Africa and Latin America had just achieved freedom. After a long colonial past they had their sovereign republics, which wanted to catch up with the western countries. Not that no contradictions existed in the process of nation building, but among the westernized, educated, nationalist elite there was a strong desire to see their country develop in the image of the industrialized countries (particularly, USA or Soviet Union), by giving a high dose to industry. Accumulation of capital, whether from savings or external sources, was central to this way of conceptualization (Chew and Denemark, 1996). This was reflected in the Second Five-Year Plan in India. This plan followed the Nehruvian model of industrial socialism and accorded more importance to big industry. Thus, development was defined as the ability of the economy to grow irreversibly at a desired rate.

Around the same time, in the advanced countries that had already achieved a high level of economic development, it became clear that like other interventions, development has its own consequences some of which make it self-limiting. Forrester (1971) and many others asserted that a framework be developed for speculation on the future. Forrester had shown how factors such as crowding, pollution, food supply, and natural resources could bring the exponential population growth to a halt, and in extreme cases they could also lead to sudden and tragic collapse of population. This inference gave rise to Malthusian credence that development not accompanied by population control, would in the long run cause misery (rise in death rate). At this point, the centre of economic discourse shifted from capital to natural resources (later called natural capital). *The Limits to Growth* (1972) prepared on the basis of the famous work of Meadows and others sponsored by the Club of Rome declared that the developmental processes are limited by their own results and thus the much-valued growth cannot be sustained. Studies by Mesavoric, Pestel (1974) and Barney (1980) established that development is neither a universal nor an irreversible process. Implicit in these studies is an idea that sustainable development is that which does not lead to shortages of renewable and non-renewable resources such as water, soil, forests, fisheries, and mineral resources at a future date.

At this moment, the concept of development became reflective. The issue was: how to deal with consequences of development, which threaten the growth of development in the future? In 1987, Brundtland Commission defined that sustainable development is that which meets the needs of the present without compromising the ability of future generations to meet their own needs. It was said that living standards that go beyond the basic minimum are sustainable only if consumption standards everywhere have regard for long-term sustainability. The concept of sustainable development acquired prominence after United Nations Conference on Environment and Development, held at Rio in 1992. It identified population .policies as an integral part of sustainable development. Principle 8 of the Rio declaration stated: "to achieve sustainable development and a higher quality of life for all people, States should reduce and eliminate unsustainable patterns of production and consumption and promote appropriate demographic policies". Since then population came to be seen as the most important factor of sustainable development.

Sustainable development is defined as the process of development, which lasts. It emphasizes the long-range development potential of environment. Thus, it suggests looking at sustainability through its impact on environment. From this perspective the direct impact on the environment may be determined by interaction of three factors: size of population, consumption standards, and technology effect on wastage of resources for each unit of consumption (UNFPA, 1992).

Impact = Population × Consumption per Person × Technology Effect

The above formula (I = PCT) indicates that since the impact arises from multiplicative relationships between population, economic development, and changes in adversity measure of economic development, there is a need to work at all levels. In other words, the concept of sustainable development suggests that efforts are made to:

- Reduce population growth rates, particularly in those regions which are passing through the second stage of demographic transition.
- Put a halt to economic development, particularly in developed countries where average consumption per person is very high. For example, Americans eat 815 billion calories of food each day—that

is roughly 200 million more than needed—enough to feed 80 million people (Internet, 2001).

- Develop new technology that yields the same levels of economic development with reduced impact on environment.

Yet, this emphasis on environment (if it means putting halt to industrialization) was not acceptable to developing countries including India where poverty rather than industrialization is conceived to be the major source of environmental degradation, and where political democratization has raised people's expectations regarding development. It should be noted that in case of less developed countries like India there are some more serious threats to sustainable development than the above equation would imply. In India, the growth rate of population has started declining but due to a strong population momentum it may remain high for quite some time. With rapid growth of population to maintain the current standards of living is becoming more difficult. Moreover, there are few efforts to control or make the best use of technology. The following sections throw light on some of the dilemmas of development in the context of India where levels of development are low from the world perspective.

Dilemma I: A strong population policy or a soft welfare approach

Population is certainly a big bottleneck in development. According to the preliminary results published by the Registrar General, India had a population of 1,027,015,247 on 1st March 2001. It showed a decadal growth of 21.34 per cent. This rate is quite alarming. Before 1921, the growth rate of India's population was somewhat erratic. After 1921, the decadal growth rate reached a level above 11 per cent. It remained at a moderate level till 1951. After 1951, all censuses have yielded a decadal growth rate above 21 per cent. The maximum rate was observed for the decade 1961-71. During this decade, the population of India increased by 24.80 per cent. Since then it has been constantly but only slowly declining. It may be noted that the present rate of natural increase for the more developed countries is as low as 0.1 per cent, and that for the less developed countries is 1.7 per cent. However, the data on vital rates show a slightly less pessimistic picture. The data show that the rate of natural increase of population of India is around 1.8 per cent (Population Reference Bureau, 2000). The birth rate of India is 27 per cent and the death rate is 9 per cent. A detailed analysis of population dynamics shows that in the future as the effect of population momentum is weakened, growth rate may be declining faster than in the past. At the same time the projections are that at least for fifty years population of India may continue to rise. According to the projections made by the Population Foundation of India (2000), during 1996-2051, the population of India is likely to grow by 76.23 per cent.

Needless to say, any policy for sustainable development in India would have to have an effective strategy to control the population growth in the country. Rapid growth of population in the present context could bring in its trail declining standards, unemployment, inequality, marginalization among the increasing number of people, social fragmentation, shortage of capital and fall in natural resources. Although it may be difficult to estimate what is really the 'carrying capacity' of India, there is little doubt that the country has already crossed that number; growth of population is bound to deteriorate the quality of population further. There is, therefore, a need to implement the National Population Policy, 2000 and bring the total fertility rate to the replacement level.

Yet a strong will to tackle the issue is lacking. The mood among the international donors as well as planners in India is to adopt a soft approach to population control. The current thinking is that the target of any intervention should be social development and justice rather than family planning. As a result of this, family planning programme is not paid enough attention and couple protection rate in some of the poor states is not reaching the desired level.

Dilemma II: Catching up with the West or sharing poverty

With a growth of population at 1.8 per cent and an acceptable capital output ratio of 4.3 per cent, investment of 7.74 per cent of national income will be just enough to maintain the present standards. With an investment of 24 per cent, a growth rate of nearly 6 per cent is possible, which will yield a growth rate of 3.3 per cent in terms of per capita income. But can the country afford to ignore investment in human capital? Some part of the investment will be required for education, health and other components of human capital. During 1960-90, the per capita income in India could grow at an annual rate of about 1.7 per cent. It is clear that India is one of the poorest countries of the world and is bound to remain so in the near future. The Finance Minister has indicated a growth rate of gross domestic product at 4.5-5.5 per cent for the year 2001-02 (Sinha, 2002). Recently, the Prime Minister has called for achieving a growth rate of 8 per cent. Let us accept that it is not possible to catch up with the western countries. As a matter of fact, at the time of independence the ratio in per capita income of India and USA was about 1:14.8. It has grown to more than 83 (India Policy Institute,

2001). This means that for the survival of civilization either we go for equitable distribution of limited wealth along with control of wants (as suggested by Gandhi), or face threat from increasingly marginalized people at a time when overall prosperity is rising (though at a slow rate).

Dilemma III: Modernization of agriculture or sustainability of resource base

One in five people in the world do not get enough to eat. Africa now produces 27 per cent less food per capita than in 1964 (PBS, 2001). After independence India has done remarkably well in agriculture: the population of India grew at a fast rate, and yet, thanks to agricultural modernization, entrepreneurship, supportive government facilities and planning, productivity as well as production per hectare could keep pace with the population growth. During 1950-51 to 1999-2000 while the population grew from nearly 360 million to 1 billion, the foodgrains production increased from 50.82 million tonnes to 199.06 million tonnes, leading to an overall improvement in food situation. According to *Economic Survey 1999-2000*, growth rate of production of foodgrains (index based) was 2.02 per cent for the period 1967-68 to 1979-80 and 3.54 for the period 1979-80 to 1989-90. On the positive side:

> Agriculture enjoys a lower ICOR [incremental capital output ratio] of 2.5 to 3 and is one of the few sectors where India enjoys an international competitive advantage. A dynamic and progressive agriculture puts purchasing power in hands of millions of people and provides maximum employment, more than any other sector in the shortest time. In fact, Indian agriculture can easily employ another 120 million people just by concentrating on developing 40 million hectares of wasteland. Modernisation of agriculture will reverse the present unending migration to the cities and thereby improve the quality of life in the country's urban areas also. India has 329 million hectares of land area. Nearly half of it is arable. India is also blessed with sunshine, rain, varied agroclimatic conditions and a farming community open to innovation (Prime Minister's Council, 2001).

On the negative side, it may be noted here that just keeping pace with population implies that the shortages already existing have continued. Data show that during 1989-90 to 1998-99, growth rate of production of foodgrains in the country was only 1.80. This indicates that during the last decade it was not possible to sustain the fruits of development in the agricultural sector. For all crops—rice, wheat, coarse cereals, total cereals, and pulses—there was a drastic reduction in production. Data also show that the production of coarse cereals has declined. Over the years, there has been a significant change in the composition of foodgrains: share of rice and wheat has increased, and the share of coarse cereals and pulses has declined. While the reduction in growth rate of cereals could be attributed to substitution effect, the failure in improving growth in pulses has been termed as "quite a setback". *Economic Survey* said:

> There are limits to increasing production through area expansion as the country has almost reached a plateau in so far as cultivable land is concerned. Hence the emphasis has to be on increasing productivity levels. The area under foodgrains has more or less remained constant at around 125 million hectares since 1970-71. However, due to aberrant weather conditions during the current year, the area is likely to go down to 123.3 million hectares.

Thus, the future requirements of the growing population will have to be met mostly from raising yield on the existing farms. This implies that if the population of India grows by 50 per cent before it stabilizes, the farmers will have to produce 50 per cent more than in 2000 just to maintain the current levels which are below the standards. Limited arable land, shrinking size of farms, and land degradation (due to high crop intensity, water logging, salinization, and misuse of fertilizers, herbicides and pesticides) make this a difficult challenge. In most parts of the country, particularly in those that have experienced the green revolution, productivity is declining, water table is going down, and more and more investment is required to maintain the current levels of productivity. There is a need to learn better water management skills from both tradition and new initiatives. Failure of monsoon this year (i.e., 2002) has taught it more clearly than ever before.

Dilemma IV: Investment or efficiency of capital

So far the overall industrial production and its sectoral components have done reasonably well. During the last decade, the peak was observed in 1995-96 when the growth rate reached a level of 12.8 per cent. After this it declined to 3.7 in 1998-99. During 1999-00, it again grew to 6.2 per cent. In the assessment of Prime Minister's Council (2001):

The Indian economy needs to target a GDP growth rate of about 12 per cent per annum in five years, to be supported by a robust industrial and service sector growth rate of around 15 per cent per annum , and agriculture growth rate of around 5 per cent per annum. On the fiscal side, the government needs to target a fiscal deficit to GDP ratio of around 3 per cent to reduce the pre-emption of funds by the government from the system and utilization of domestic savings for more productive asset creation. The target for inflation should be set at around 6 per cent.

To achieve a GDP of 12 per cent per annum, India has two options:

1. Achieve and maintain investment levels at 48 per cent of GDP with an incremental capital output ratio (ICOR) of about 4.
2. Lower ICOR to the 2.5 to 3 range (through greater productivity and efficiency of capital) and mandate investment level at 30 per cent of GDP.

Obviously it is not possible to achieve an investment level of 48 per cent or an ICOR of 2.5 to 3. A plan for the future should make a realistic assessment of the economic parameters and develop appropriate policies for development.

Dilemma V: Greater industrialization or quality of environment

Environment is vital for sustainable development. It consists of air, water and land (including forest cover). Among them water is of great importance. UNFPA's report *The State of World Population 2001* says: "Water may be the resource that defines the limits of sustainable development. It has no substitute, and the balance between humanity's demands and the quantity available is already precarious." The regions, which receive higher rainfall, have higher income and are more developed than the regions, which receive less rainfall. Everywhere due to increased use of agriculture and industrial development, water use has grown at considerably higher rate than the population. Already an estimated 1.7 billion people lack access to clean drinking water, and 80 per cent of the urban dwellers are without safe water and sanitation facilities (UNFPA, 2001). According to *The State of World Population 2001*, 54 per cent of the annual available water is being used. "If consumption per person remains steady, by 2025 we could be using 70 per cent of the total because of population growth alone." In India too the story would not be very different. Water is going to be the biggest problem of sustainable development. "The government is failing to protect rivers." "Overextraction of water for irrigation and urban consumption leaves several rivers dry for most of their course, denying dilution of pollutants flowing into the river. The Yamuna has almost no water downstream of Tajewala as all of it is abstracted for irrigation in Uttar Pradesh and Haryana" (Centre for Science and Environment, 2001).

There are other environmental issues also that are quite complex. They include the rise in industrial emissions and emission of greenhouse gases, growing monoculture, i.e., greater dependence on a three to four crops, climate change, disappearance of forests, decline in global diversity, urbanization, and air quality (UNFPA, 2001). "Each year, an estimated 27,000 species of animals, plants, fungi and micro-organisms become extinct, taking their ecological services and genetic secrets with them." (UNFPA, 2001).

The Government of India has passed The Air (Prevention and Control of Pollution) Act, 1981 for the prevention and control of pollution. It is yet to be supported by social institutions. There is a need to collect data on India's environment and take corrective actions. Researchers at Centre for Science and Environment have made a good contribution in this respect. They have shown the status of environment with respect to land, forest, pollution, and dams, and studied and documented people's movements and success stories. According to *State of India's Environment: The Citizens' Fifth Report*, India's urban centres are becoming lethal gas chambers. Most of the air quality standards in India are much above than those set by the World Health Organization (WHO). Vehicular air pollution is the major culprit for foul air in urban India. Moreover, emission standards in India are so less that in the year 2000 the country adopted standards Europe had enforced way back in 1992-93. High pollution levels would cause the rise in morbidity and mortality rate and change the climate of the affected regions.

Dilemma VI: Caring for world market or for the poor

In the light of the realistic assessment of India's resources and strengths, there is a need to look at the development paradigm. Should India adopt the capital-intensive strategies used in today's developed countries or adopt a new paradigm that is capital-saving and labour-intensive with capacity to compete in the world market? On the one hand, there seems to be no way out to withdraw from the world market, but on the other hand the capitalistic paradigm cannot make India raise the economic levels for all. Also, within the country there are regional and urban-rural contradictions. The problem of paradigmatic choice is perhaps most vivid in the controversy about big dams. *The State of World Population, 2001* reports: "The construction of large dams has slowed down, particularly in more developed countries, as their disadvantages are appreciated: environmental disruption, displacement of long-settled populations, loss of agricultural land, silting and denial of water to downstream areas, sometimes in other countries. Large dam projects continue in Turkey, China and India." There is no doubt that in the postmodern age there are no clear-cut choices and the world situation has become more complex (hyperreal), but at the same time it must be repeatedly said that the paradigm of development and modernization may destroy the country. Search for sustainability is a search for new and creative alternatives.

Dilemma VII: Adopting technology or developing an environment friendly technology

In the formula $I = PCT$ given above, T has a vital role. A more favourable technology effect would go a long way to counter the effects of P and C.

To quote:

> All development projects utilize natural resources in one form or the other and also generate wastes. The quantum of wastes generated depends on the level of technology and management skills used. Obsolete technologies result in wastage of resources to the extent of 35-40 per cent. It is, therefore, necessary to develop alternate strategies for sustainable development in the context of a developing country like India (Ministry of Environment & Forests, 2001).

But the question is: whether such a technology is developed? Whether India is able to develop technology that gives her a competitive advantage in the world market, solve the problem of poverty, raise the agricultural and industrial production to satisfy the requirements of the growing population, and also contribute to environmental quality? At the moment the answer seems to be "no". It has to be changed to "yes". According to the Ninth Plan Draft 1997-2002, there should be greater emphasis on clean and eco-friendly technologies and focus should be on the concept that one industry's pollutant is another industry's raw material. It talks about "technology with zero toxicity and zero environmental impact for sustainable development," "new energy efficient and environment-friendly processes based on automation and artificial intelligence, in addition to those concerning fiscal and material resources," "Integrated Mission for Sustainable Development (IMSD) of the DOS," and "to preserve, conserve and protect the marine flora and fauna and to promote sustainable development through proper use of biological resources". Thus, the government is committed to sustainable development but the fruits of planning are yet to come.

The Government of India is already engaged in programmes for hazardous wastes, forest cover, air quality, river water quality, afforestation, wasteland development, and species protection. Several voluntary organizations are also involved. But, until the people at the grass roots and the entrepreneurs have developed a strong environmental consciousness and have to pay for the deterioration of environmental quality there may not be much success on this account. There is a need to develop technology for action at all levels—big and small—that leads to sustainable development.

Dilemma VIII: Focus on economic development or social development

There are other socio-cultural and political impediments to sustainable development. At present the country is marked by a high degree of inequality. As many as 34 per cent people are living below the poverty line (Sundaram, 2001), while some are enjoying as much wealth as people in the developed countries. "Poorer people, who cannot meet their subsistence needs through purchase, are forced to use common property resources such as forests for fuel, pasture for fodder, and ponds and rivers for water. Population growth in poverty conditions further increases the pressure on natural resources" (Gulati, 2000). Consensus and strategies about future are unlikely to succeed without a consensus on social justice. In absence of justice the society remains highly divided, breeds suspicion, weakens social integration, and in extreme cases promotes secessionism and terrorism. Why should a dispossessed or deprived, worry about the future generations of the affluent class? To the authors, the conflict between the present of the poor and the future of the presently rich is the most important question of the sustainable development in India. Globalization, recession, rising unemployment are bound to promote cleavages and inequalities. Whether India would make any development in the future will depend heavily on whether she is able to secure participation of all in the emerging patterns of development.

One of the most crucial issues of social policy is health. The country is far behind the developed countries in conventional measures of health, such as infant and child mortality rates, maternal mortality rate, life expectancy, and morbidity patterns. New and more serious risks have come into being. For example, India is facing a serious risk in the form of

Acquired Immuno-Deficiency Syndrome (AIDS), the late clinical stage of Human Immuno-Deficiency Virus (HIV), recognized only recently. Data on estimates of prevalence of HIV/AIDS show that HIV incidence in India is moderate and increasing. In India, HIV prevalence is low but increasing and there are significant regional variations in it. There are already localized epidemics within high-risk groups in India, and the virus is spreading to the general population. Studies show that more than 3 per cent women in Bombay were HIV-positive. So far the dominant perception of society has been that HIV cases are confined to specific groups such as: commercial sex workers and their clients, migrant workers in metropolitan cities, injection drug users and vulnerable women (wives who cannot protect themselves and who cannot negotiate for safer sex). When the literature on the subject is scanned, it is found that HIV virus has started spreading to rural areas, and the general population. It may be noted that the demographic impact of AIDS is much higher than what the prevalence rates would suggest. Moreover, the current trends in HIV infection will have an impact on rates of infant, child and adult mortality, life expectancy and economic growth in many countries (Anonymous, 2000).

At one time, it was thought, as during the Fourth Five-Year Plan in India, that the fruits of development should reach all people. Experience has shown that equality is the very condition for development. Fragmentation of society, vertical or horizontal, is a great threat to development. Peace is a necessary condition of development. Inequality, divisive politics, social and economic contradictions are factors that generate secessionism, terrorism, corruption, inefficiency and aggression. For sustainable development needs of the downtrodden and dispossessed would have to be taken care of. This has to be achieved through increasing democratisation of public space, and consensus and participatory processes at the grassroots level. There is no alternative to emancipatory action.

Concluding Remarks

In this paper the authors have adopted a broader approach to sustainable development. In principle, it includes all the socio-economic and policy issues that could stop development and put a large population to survival risks. It has identified eight dilemmas of development. It must be stressed that failure to evolve a pattern of sustainable development may not simply cause de-development but may lead to a disastrous form of development. In development there is no going back. Reversal would not mean going back to a state of past but unprecedented immiseration and violence. This is the time when state and civil society should learn from the reflections on development in the developed countries as well as the less developed countries, and work at all levels to achieve and manage sustainable and acceptable levels of development. It is not going to be a simple task. It requires not only working out some kind of optima in the given world situation but it will also require new conceptual apparatuses in the light of which sustainability can be defined and practiced: it requires learning from the scientific analysis of the dynamics of the world economic system as well as perspectival understanding from the grassroots initiatives.

References

Agarwal, Anil and Sunita Narain, *Dying Wisdom*, "State of India's Environment: A Citizens' Report", Centre for Science and Environment, New Delhi.

Agarwal, Anil, Sunita Narain and Indira Khurana (eds.) (2001), *Making Water Everyone's Business: Practice and Policy of Water Harvesting*, Centre for Science and Environment, New Delhi.

Anonymous (2000), "AIDS will Kill One Third of 15-Year-Olds in Worst-Affected States", *Populi*, 27:2, 2000.

Barney, G.O. (1980), *Global 2000: The Report to the President, Entering the 21ˢᵗ Century*, Washington D.C.: U.S. Government Printing Office.

Centre for Science and Environment 2001 *http://www.cseindia.org/html/extra/soe/soe5_hl.htm*

Chew, Sing C. and Denemark, Robert A. (1996), "On Development and Underdevelopment," in Chew, Sing C. and Denemark, Robert A. (eds.), *The Underdevelopment of Development*, London: Sage Publications.

Forrester, J.W. (1971), *World Dynamics*, Cambridge: Wright-Allen Press.

Gulati, S.C. (2000) "Population-Development-Environment Linkages in India," *Millennium Conference on Population, Development and Environment Nexus*, IASP, PFI and UNFPA, Ph. D. House: New Delhi.

India Policy Institute *http://www.indiapolicy.org/debate/Notes/data1.html*.

Ministry of Environment & Forests 2001 *Annual Report 2000-2001*, Government of India, http://envfor.nic.in/report/report.html

Prime Minister's Council on Trade and Industry (2001), "Reforms in the Financial Sector and Capital Markets: A Vision for India," http://www.nic.in/pmcouncils/reports/fin/vision.htm

Meadows, D. H., J.R. Meadows and W.W. Berhrens III (1972), *The Limits to Growth*, New York: Universe Books.

Mesavoric, M. and E. Pestel (1974), *Mankind at the Turning Point*. New York: E.P. Dutton.

Natarajan, K.S. and V. Jayachandran (2000), "Population Growth in 21st Century-India," *Millennium Conference on Population, Development and Environment Nexus*, IASP, PFI and UNFPA, Ph.D. House, New Delhi.

PBS http://www.pbs.org/kqed/population_bomb/danger/price.html

Population Reference Bureau, *World Population Data Sheet*.

Sinha, Yashwant (2002), Agency report in *Dainik Jagaran*, 6 January.

Sundaram, K. (2001), "Employment and Poverty in 1990s", *Economic and Political Weekly*, 36(32): 3039-49.

United Nations Population Fund (1992), *The State of World Population*, New York: UNFPA.

United Nations Population Fund (2001), *The State of World Population*, New York: UNFPA.

World Commission on Environment and Development (1987), *Our Common Future*, Oxford: Oxford University Press.

24

Development, Environmental Degradation and Marginalization of Women in South Asia

Vandana Asthana

The term 'ecofeminism', as coined by Françoise d' Eaubonne in 1974, means many things, but has at its core the notion that women have a unique relationship to nature, ground in their intuitive ethic of caring and preserving. Inherent in this thinking is the idea that women have the potential to bring about an ecological revolution to save the planet. Some have called it a movement, others a discourse. In its early inception, since the 1970s, it was a movement of mostly western women to reclaim a spiritual relationship with the earth, connecting the life support system of nature with women's innate life support systems of mothering and nurturing beings, in understanding the relationship with the natural world. This world of "a nurturance and close human relationship is the sphere where basic human needs are anchored and where models for humane alternatives can be found. This world, which has been carried forward mainly by women, is an existing alternative culture, a source of ideas and values for shaping an alternative path of development for nations and all humanity".[1]

The early period of ecofeminism celebrated woman's emotional connection with the earth forged through her biological functions, characterizing it as distinct to the reason-dominated approach associated with men, empiricism, and power. The movement sought to accomplish two goals: to establish the interconnections between feminism and ecology, and to demonstrate the inadequacies of environmental theory for accommodating insights of women.[2] In the later periods, led by the work of Vandana Shiva, a southern women's perspective of ecofeminism evolved, linking experiences of colonialism and western development to women's special relationship with the earth.

While there are different approaches to understanding women's inherent relationship to nature the ecofeminist discourse relates these differing cultural and economic perspectives to the notion that both women and nature have been degraded, dominated, and devalued by western culture. The ecofeminist discourse has evolved to reflect the periods of the environmental movement. It interfaces with the issues championed by the women, environment and development movement by challenging the dominant models of environmental protection characteristic of the West and patriarchal domination of both nature and women where the role of the male in reproduction was elevated above the role of the female. There also began a shift from Goddess-centred cultures to male deities and woman was created from Adam's rib and placed below him, and below these two were the animals and the rest of nature, all to serve man. Advocacy of ecofeminism has been growing, especially, in the West but apart from some third world variations; it contemplates the link between women and the environment mainly in ideological terms. The basic line of argument is the commonality of important interconnections between the domination and oppression of women and the domination and exploitation of nature. There has emerged a global shift in orientation, from reverence and respect for women and nature, and an interconnected perspective of life, to a worldview which is based on separation, and which conceptualizes women, nature, and animals as inferior and subordinate objects to be dominated.

While the current system of global inequity, international violence, and environmental degradation may seem beyond the scope of feminist analysis initially, the debate over gender dimensions *vis-à-vis* environment and development intensified in the 1980s and came to be recognized the world over since the Rio Earth Summit in 1992. Even after two decades since the United Nations recognized the need to address women's problems globally, the question of women's dependence on natural resources still remains valid. It was at the "Forward Looking Strategies of Nairobi Conference 1985" that the link between productivity of women and environment first drew attention. It was emphasized: "Deprivation of traditional means of livelihood is most often a result of environmental degradation resulting from such natural and man-made disasters as draughts, flood, hurricane, erosions, desertification, deforestation and inappropriate land use. Such conditions have already pushed great numbers of poor women into marginal environments where critically low levels of water supplies, shortages of fuel, overutilization of grazing and arable land and population density have deprived them of their livelihood. Most seriously affected are the women in arid and semi-arid areas, in urban slums, and in squatter settlements. These women need options for alternative means of livelihood, improvement in home and work environments and enhanced capacity to manage their environment and sustain productive resources."[3] In spite of this perceptive understanding nearly a decade ago, there are many, who still query that how is environment a specific issue of women? Environment degradation of land, water, air and the consequent imperative of conservation of the same is perceived as a general problem but nothing specific to women. A few assumptions, therefore, need to be stated at the outset:

(i) It is not the socialite, high and middle class women that we refer to here, but of the millions and millions fighting battles of survival itself in hills, deserts, marginal degraded lands, flooded plains, coastal areas and in slums and squatter settlements of our metropolitan and mega cities all over the world.

(ii) These women do not form a separate group as such, but belong to that vast mass of humanity referred to as the 'bottom billions' in the UN and other documents. Women and the girl children are the most deprived of this group of poor people and so the greatest sufferers with respect to all basic minimum human needs.

(iii) Attempts at redemption within the policy frame have to be directed at the whole group but specifically targeted at women and the girl child.[4]

For those who live outside of the wealthy industrialized world, environmental degradation has immediate, tangible results—hunger, thirst and fuel scarcity. It is only the majority of the world's less privileged, especially women, children and people of colour that are impacted the most.

The difference between perception of women and men of the difficult reality facing them lies basically ascribed different general roles, behavioural patterns, attitudes and attributes to them. "Women have brought a different perspective to the environment debate. Because of experience base, poor women's lives are not compartmentalized and their work is not seasonal. They, therefore, see the problem from a much broader and more holistic perspective. They understand more clearly than policy makers that economics and the environment are compatible. Their experience makes this clear to them, because the soil, water and vegetation, which the poor require for their basic livelihoods, need specific care and good management. Women from the South do not separate people from the natural resource base."[5]

Thus, while the scientific community, from the seventies onwards, initially visualized environment as a problem of physical pollution and degradation of the bio-sphere, the suffering and deprived humanity and women find it to be a problem of the socio-economic sphere. The problem lies in the inequitable international appropriation and distribution of the resources of the world in which the gender relations make women and the girl child the most deprived, really decimated in fact, poor and powerless.

"Contemporary development activity in the Third World superimposes the scientific and economic paradigms created by western gender-based ideology on other communities. Ecological destruction and the marginalization of women, we know now, have been the inevitable results of most development programmes and projects based on such paradigms; they violate the integrity of one and destroy the productivity of the other. Women, as victims of the violence of patriarchal forms of development, have risen against it to protect nature and preserve their survival and sustenance. Indian women have been in the forefront of ecological struggles to conserve forests, land and water. They have challenged the western concept of nature as an object of exploitation and have protected her as 'Prakriti', the living force that supports life. They have challenged the western concept of economics as production of profits and capital accumulation with their own concept of economics as production of sustenance and need satisfaction."[6]

Role of Women

It is difficult to talk about women as a whole without ignoring the vast economic, cultural and social differences between them. Even if we were to consider only Third World women, the lives of women in South Asia are different from those of African or Latin America and within countries too, similar gaps in income and culture exist. Nevertheless, certain commonalities shape the women's living conditions in the rural areas of the third world in general and South Asia in particular for example, poverty. Roughly, 75 per cent of the world's population is among the poorest, and women make up the majority of the poor. Wherever they live, they are bound together by the common fact of their tremendous work burden. Thus, rural women have traditionally been the invisible workforces, the unacknowledged backbone of the family economy. The crucial areas where women work to keep the family and rural economy alive can be identified as

1. survival tasks,
2. household tasks, and
3. income-generating tasks.

Survival Tasks

Survival tasks constitute those essential for daily life for which women are primarily responsible. They grow the food crops, provide water, gather fuel and perform most of the other works, which sustain the family. A certain division of labour is evident in the agricultural sector. Women are generally responsible for sowing, weeding, crop maintenance and harvesting as long as these tasks have not been mechanized. Men on the other hand look after field preparation while subsistence agriculture, i.e., growing of food crops is the task of the women. Presently, African women perform 60 per cent of the agricultural work and 60 to 80 per cent of the food production work.[7] Her participation in growing cash crops is also increasing but while they work harder than men they receive no compensation.

Moreover, cash crops had led to draining of the natural resources of the developing countries. The replacement of mixed tropical forests by non-native cash crops of eucalyptus trees and sugarcane in India has led to deforestation and water loss. The resultant deforestation and water loss has meant longer and longer walks each day for rural women to gather fuel and to haul water.[8] The supply of water is crucial to their survival and health of the family. Women walking long distances to collect water consume sums of total calorific intake up to 12 to 27 per cent. Apart from nutritional loss, it is also time consuming and a tiresome job.

For their energy supplies, the rural areas of the South Asian region depend mainly on biomass such as fuel wood, crop residues and manure. 75 per cent of the rural energy supplies come from biomass. Fuel collection of women is facilitated with the help of children. Depending on the ecological characteristics of the area in which they live, women may spend up to five hours a day on fuel collection. Women in Delhi have to walk an average of 10 kilometres every three out of four days for an average of seven hours at a time[9], just to obtain firewood bearing the disproportionate burden of underdevelopment.

Household Tasks

The activities of the home are exclusively the responsibility of the women with older children occasionally assisting. The daily tasks including food preparation, cleaning and washing absorb long hours of work daily. Globally, women produce approximately 80 per cent of the world's food supplies, and for this reason women are the most seriously affected by such food and fuel shortages. While women produce the food, the men and boys are served first and provided with the most nutritious foods.[10] Women in many cultures are the last to eat in the family.

Income-Generating Tasks

Throughout South Asia women contribute substantially to the family budget through income generating activities like food processing, trading of agricultural products and handicrafts production. Such kinds of activities are found in both situations where men have migrated to cities or abroad and in households where women are the sole providers. The contribution of the women to the budget is of utmost importance to the family as women spend more of their income on family welfare.

Apart from gathering, sowing, cooking, washing clothes and more, these are tasks of paramount importance performed by millions of women every single day of their lives. Despite some progress in correcting unequal gender relations, such tasks primarily remain "women's work". Over the last decade or so, nationally and internationally, the reality that deterioration in the natural environment has a direct impact on women's lives has been accepted.

Furthermore, technological innovations have rarely addressed the need to lighten the workload on women, particularly rural poor women. Millions of women spend hours in inefficient, smoky cooking stoves, which are deleterious to their health. This is a factor that has never been considered an important problem. Despite advances in agriculture, women still have to perform the backbreaking tasks of weeding, transplanting, harvesting and processing. Thus, women fulfil a great number of essential tasks, yet they and their labour go unrewarded. "Although women represent half of the world's population and one third of the official labour workforce they receive only one per cent of the world income and own less than one per cent of the world's property"[11] and despite massive involvement of women socially and environmentally her role and contribution remains lesser acknowledged than it deserves.

The United Nation's System of National Accounts has no method for accounting for nature's own production and destruction till goods do not reach the market economy, nor does it account for the majority of the work done by women. Forests, fresh water supplies that provide fuel, food and fodder to women have no economic value in the UNSNA system unless they can be traded as commodities in the market. Thus, women and nature do not count in the international market economy.

Women and Environment Interweb: Current Trends in South Asia

In the recent decades, two distinct but interactive macro-processes have severely affected the South Asian rural environment: the depletion of its natural resource base and the increasing appropriation of what is available for the benefit of a few.[12]

The problem, challenges and issues seeking women assistance are multifaceted from deforestation, land degradation, floods, droughts, pollution, burgeoning population, poverty, loss of biodiversity, adverse climate change, etc., Most of these issues bridle down to the basic issue of peoples' control in general and women's in particular over basic resources and livelihood patterns, effective role in decision-making and empowerment.

All ancient river-valley civilizations including the Indus valley civilization were based on women's discovery of primitive agriculture. The reproductive creativity of women and the productive fertility of mother-earth, through women's intercession, were imposed on each other in societies based on primitive hoe-agriculture through magico-religious fertility cults. It is this substratum which gave rise to the great 'Tantra' cults and 'Samkhya' philosophy which regards 'Prakriti' (nature), envisaged woman as 'Annapurna' (full of cereals) and the practice of Goddess worship as 'Jagdishwari' (the Master of the Universe), and 'Jagatmata' (the Mother of the Universe). With the taking over of hoe-agriculture by plough-bullock agriculture in the Indo-Gangetic plains, after the dense forests were cleared, men's role in agriculture increased. Still women continue to be the unrecognized farmers. However, it needs to be noted that there is no word coined either in Sanskrit or Hindi for women farmers, again symptomatic of the Aryan patriarch's blindness to the indigenous women's roles in agriculture. Forests and agricultural plains have an intimate relationship. Forests usually act as catchments for rivers. To harness water in order to increase food security, irrigation projects were taken up from time to time.

"Typically, these would submerge forests, displace local communities, and benefit agriculture in the plains. Today some of these, like the dam on the river Narmada and the Tehri dam, have become areas of serious debate. This is an issue which posses very difficult questions regarding the environment and development challenge displacing and ousting women."[13]

Existent depletion and disappearance of forests of South Asia, deteriorating soil conditions and declining water resources call for women to take up the challenge. Half of potentially irrigable and culti-vable area is lost due to developmental projects and water logging. The availability of ground water is also falling. Fertilizers and pesticides runoffs have polluted many natural water sources making them unfit for aquatic life and human role. Alongside this depletion, is the progressive state ownership and privatization leading to appropriation of natural resources, for the benefit of a few. The earlier rights—community collection and use, irrespective of gender are eroded. Poor households in particular have depended on lands for basic necessities which until recently in many states of India were supplied by Village Commons (VCs), 90 per cent of fuel wood and 70 per cent of the grazing needs of the poor, compared with the relative self sufficiency of larger landed households.[14] Forests and water are other forms of common property resources providing subsistence provisioning for the poor.

The area of VCs had declined dramatically in most countries of South Asia with illegal encroachments and government distribution of land to individuals under various land reforms and poverty removal programmes. Aimed to help the poor, the land has benefitted the better of people in many regions. In India, over 80 per cent has gone to the already landed and the poor have lost collectively while gaining little individually. The productivity of what remains is also declining under population growth and unmonitored use. Similar trends are also visible in Pakistan.

This process of state ownership and privatization has not only concentrated natural resources in the hands of a few; they have also contributed to their depletion by undermining the traditional institu-tional arrangements of communal resource use and its management which existed in many regions. Such traditional systems of water management, grazing and firewood gathering were typically not destructive of nature. But government takeovers and privatization have made these a commodity to be individually exploited and commercially used; the effects of which is greatly felt by women.

Biomass

Biomass plays a crucial role in meeting daily survival needs of the vast majority of South Asian population and rural households of the region. Food, fish, fuelwood fodder, fertilizer (conducing, organic manure, forest litter) building material (timber, thatch) and medicines (herb) are all different forms of biomass. Water is another crucial element and its availability is definitely related to biomass in a country where there is rainfall for only four months in the year. When the biomass in the surrounding environment disappears, water sources like streams and ponds also tend to disappear soon after the monsoon. An important aspect of this biomass-based subsistence economy is that it is mostly non-monetized. Water, fuel, fodder, building materials and even food to a certain extent are all gathered freely from the immediate environment: production and processing of biomass—agriculture, forestry, minor forest produce and village crafts based on biomass as raw material—are also the biggest sources of employment.

In the name of plantations, extended and improved agriculture, irrigation, power generation, means of communication, industrialization, monetization and mono-culturalization of the agriculture and urbanization, i.e., development imperatives for the good of the well-to-do in the name of the nation biomass has been destroyed on a vast scale through deforestation and devegetation.

This destruction of biomass, has adversely affected women the most, particularly from landless and marginal farming families in the rural areas, tribal women from denuded forest areas, nomadic women for overgrazed lands, women in coastal areas depending on fishery and migratory women in slums and squatter settlements. Traditionally women have been responsible, in these subsistence or survival economies for water, food, fuel, fodder, habitat, etc., that are the survival needs of the family. Women's symbolic relationship with nature has been recognized since ancient times through the concept of 'Van Devi' (forest goddess) and sacred groves in India's cultural traditions. Their rearing and caring roles for children have been extended to nature also, which sustains their existence in a mutually supportive relationship. They, therefore, have been the greatest conservationists.

Government control over forests has definitely meant a reallocation of forest resources away from the needs of local communities and into the hands of urban and industrial India. The end-result is both increased social conflict and increased destruction of transformation of the ecological resource itself. There is, however, a general recognition now that the destruction of their natural environment and resource base and prevention of access of biomass not only makes their own, and family's survival difficult but this alienation at times forces them involuntarily into non-conservationist roles. A synoptic look at different sectors of ecologically degraded spaces would illustrate this better, on the basis of studies made over the last decade in India.

Deforestation

While colonial rulers treated the forests as 'timber mines', extensive deforestation has been taking place in the service of development, urbanization and international markets. While conservation and regeneration of forests come as a priority in South Asia policy making and implementation, serious conflicts often occur between gangs out to denude forests and conservation process of forest departments. As long as forests were linked to community management, responsibility for resource management was linked to resource use. But government control and privatization seem to have eroded the social base and broken the link of this protective system. Women are more dependent on forests and thus more affected by forest destruction as compared to men, due to multiple responsibilities of gathering firewood, fodder and food for households. Field surveys often reveal that women's needs often receive no consideration. Women remain marginalized at the community level as well as at national policy framework because of their disadvantageous position ascribed by patriarchal gender relations, traditionally excluding women from political participation, even at the community level.

The traditional responsibilities of man-women relationship in forest management influences their perceptions and priorities. In most tribal and non-tribal communities living in the forest areas in central and eastern India, men are normally responsible for providing timber for construction of house and agricultural implements and women for the daily supply of firewood, fodder and water. There seems little objectivity in fixed gender division of the above roles as they are known to change dramatically by caste, tribe and class within the same village but the general pattern remains identical.

Tribal Women and Deforestation

The tribal communities today find themselves thrown out of their traditional habitat. They had owned forests collectively, empowering them rights of collection of dried wood, branches, grasses, minor forest produce, etc. With the State Forest Corporations taking over, their hereditary rights are denied. Under stress of land and its availability men migrate, and women are left to fend for themselves and the children. What was rightfully theirs and sustained them, has been taken away and their attempts at access are transformed into poaching, stealing and so punishable by law also. Two to three million people in the forests and tribal belt of middle India from Ranchi to Gwalior, earn their livelihood by head loading, 90 per cent of whom are women. Harassed, robbed and exploited by officials and timber merchants, they trudge miles to sell these loads of wood at a pittance and begin the grind every alternate day. In spite of the backbreaking nature of this work, women during the last two decades, have taken it up in large numbers because they cannot rely on men to handle cash, and also because this work is available all the year round when no other work is available. With mothers being out working all the time, the girl child has no other option left but to act as a foster mother to her siblings and has to be willingly or unwillingly deprived of all opportunities to become a productive member of a fast changing and modernizing economy requiring skilled labour. And so being deprived of education and technical training at different levels, not only the present generation of women but future generations to come are thus being deprived of a fundamental right of equal opportunity guaranteed by the Constitution of India. Similar instances could be cited from other South Asian states.

Collection of minor forest produce and herbs not only provided extra income to women but valuable nutrition and medicinal support. Denial of this has made millions into nutrition-poor without providing any alternative. Living in a labour-intensive and demonetized economy they have been suddenly, in the last few decades, exposed to the vagaries of cash economy and become the double victims of profiteers, middlemen and also their own men. Cash is usually handled by men and often spent on non-essentials. Research has found that women's income at this level mainly fulfils household and nutritional needs of the family.

Pastoral and Nomadic Women

Provision of fodder and taking care of livestock traditionally has been women's responsibility. With pasturelands degraded, forests receding, cash economy and biogas plants utilizing dung and even agricultural wastes, the condition of landless women is worsening. Many have given up their hereditary vocation and have joined the large and growing army of landless labour in government food-for-work or draught relief programmes. The increased number of women in these programmes is only symptomatic of the fodder crisis in the country. Women who are left behind are forced to dig up even roots of trees and other vegetation from the parched land both for cooking and other purposes. Water scarcity in these avid areas forces them to walk 2 to 3 kilometres to fetch water for drinking, cooking and household purposes. As croplands extend into grazing lands, grazing extends into forests, thus giving rise to conflicts between the grazers and foresters. As the nomadic grazers from other states passed by the settled fields, earlier the farmers welcomed them for the manure that the cow hands left, and even paid them. With cash cropping and increasing use of chemical fertilizers this synergic ecological relationship between the nomadic grazers and settled farmers has changed into one of conflict. Thus, the degradation of the environment has put a greater pressure on the women below poverty line in these communities.

Fisherwomen, Technological Inroads and Depletion of Marine and Riverine Resources

South Asia has major riverine systems with communities. In fisheries, women are mainly involved in all processes after the fish is landed. The main operations in which women are involved include fish vending, fish processing, drying and curing, prawn processing, loading and unloading and pot making. Apart from the usual difficulties of house-to-house vending, women here face the usual rigours of a male dominated market. In spite of huge increase in the production of both marine and riverine fish and prawn and great export potential, women face special difficulties in processing and marketing due to extremely low wages, sexual and other harassment of younger and unmarried young women, through unlicensed contractor managed shelling and grading sheds and landing places.

Implications of Modern Technology

Modern technology for increasing agricultural output has also had high environmental costs. The Green Revolution technology, in particular, with its high dependence on chemical inputs and an assured water supply, while dramatically increasing output in the short term, has over time led to falling water tables due to the indiscriminate sinking of tubewells, waterlogged and saline soils from many large irrigation schemes, declining soil fertility with excessive fertilizer use, water pollution with pesticides, the loss of genetic variety with monoculture cultivation and the marginalization of indigenous knowledge systems.

Population growth has impinged on these processes more as an exacerbating cause than as a primary cause of degradation. Commercial exploitation and the appropriation of forests and VCs by a few, pushes the vast majority for subsistence on depleting resources, causing their further decline; the poor, in particular, who depend on these resources, may be seen here as victims of the crisis not its causes.

Finally, the high energy-intensive consumption patterns of the elite, the types of products consumed and the technologies used to produce them, have all contributed to the pace of environmental degradation.

Another contradiction arises from the very success of the Green Revolution since the sixties, because the environmental costs were never taken into account. Soil health problems, deficiency of micro-nutrients, salinity of soil and the damage to the physical structure of the soil and rising demand of chemical fertilizers have all led today to a situation which is not very happy.

Both humans and animals suffer from the introduction of scientific manipulation into production and reproduction. Near 143 mha of agricultural land out of the 266 mha, which is potentially productive, 85 mha suffers today from varying degrees of degradation. That in spite of the vast irrigation potential, wastelands have grown is not only a travesty of development but symptomatic of blindness to environmental costs. The increase in biological manipulation in agriculture and growth of the biotech industry are of concern to ecofeminists. The social and political planning that went into the Green Revolution aimed at engineering not just seeds but social relations as well.[15]

More than 40 per cent of India's population lives in the Indo-Gangetic plain, and it is here that soil health is getting worse. The rising urbanization of India as documented by the 1991 census, 27 per cent in fact, has been due to 30 per cent to 40 per cent migrants from the rural/tribal/hill areas. With men migrating first in search of livelihood, the number of women-headed households has gone up to 20 per cent to 35 per cent and these, as is the case world over, are the poorest households. The large number of women in food-for-work and drought relief works are a testimony of the burden women face in somehow scraping a living to support the children as best as they can. However, a little more than two thirds of India's workforce and nearly 84 per cent of all economically active women are engaged in agriculture and allied activities.

Gender-Specific Effects of Degradation

These processes do not affect everyone equally. It is the poor, the landless households that are most seriously affected especially those located in high-risk areas such as hills and semi-arid plains. Within households the negative effects are borne disproportionately by women and female children. The gender-specific effects arise especially from pre-existing inequalities, notably:

An unequal gender division of labour: Women and female children do much of the gathering and fetching from forests, VCs, rivers and wells, and poor peasant women's daily work routine can total 12 to 15 hours, typically many hours more than worked by men.

Gender inequalities in the intra-household distribution of available resources: There is a systematic anti-female bias in access to healthcare and to some extent also food within rural families, especially, (but not only) in northern India, revealed by the range of indicators, but most starkly in sex ratios (females per 1,000 males). These are female adverse in virtually all of India, except Kerala. In the absence of a bias, given the biological longevity of women, we would expect there to be more women than men in the population, as in most parts of the world.

Women also have fewer earning opportunities, enjoy lesser job-search mobility and typically receive lower pay for the same or similar work. Given women's limited access to private property resources, to cash and to marketed goods, their dependence on common property resources has always been much more substantial than that of men of the same households.

Women's unequal access to knowledge systems predicated on modern science and technology and a low valuation of their traditional knowledge systems; women's unequal access to decision-making authority at all levels, including decisions about natural resources, etc.[16]

Invisibility of Women's Greater Work Burden and Longer Hours

Agriculture policy, in spite of the National Perspective Plan (1988), still treats women only as housewives and not farmers. Women's work is invisible because of the inability of statisticians and researchers to define their work in and out of the house and farming is usually part of both. It is also invisible because "women are concentrated outside market-related or remunerated work, and they are normally concentrated in multiple tasks. Studies prove that women in India are major producers of food in terms of value, volume and hours worked".[17]

More than two thirds of the labour input is female, and whether it is shifting cultivation, subsistence and low input agriculture, or high external input agriculture, women work longer and harder than men. According to the assessment by Singh (1987), in the Indian Himalayas, a pair of bullocks works for 1,064 hours a man for 1,212 hours and a woman for 3,485 hours in a year on a one-hectare farm. A women's work is more than of men and farm animals combined. For a one hectare farm women average 640 hours for intercultural operations like weeding, 384 hours for irrigation, 650 hours for transporting organic manure and transferring it to the field, 557 hours for seed sowing and 984 hours for harvesting and threshing."[18] In Himachal, in terms of total, farm work women contributed 61 per cent of the total labour.

Women's work and livelihoods in subsistence agriculture, for example, are based on multiple use and management of biomass for fodder, fertilizer, food and fuel. The collection of fodder from the forest is part of the process of transferring fertility for crop production and managing soil and water stability. Is the work of the women engaged in such activity to be counted in the forestry sector? The tendency is to leave it out from all sectors, and make it invisible. It is in the 'in between' spaces, the interstices of 'sectors', the invisible ecological flows between sectors, that women's work and knowledge in agriculture is uniquely found and it is through these linkages that ecological stability and sustainability and productivity under resource scare conditions are maintained.[19]

Consequences of Environmental Degradation on Women

On Economics of Women

The decline in forests and VCs has reduced women's incomes from gathered items and affected cattle-dependent livelihoods. In addition, the extra hours needed for gathering have cut into crop production time, affecting crop incomes, especially in hill communities where, with high male migration, women are often the primary cultivators. Thus, the cost of economic opportunities lost is very high because of the time, labour and human energy lost in meeting the basic household needs of the family.

Nutritional Implications for Women

Studies in several parts of India reveal a gender bias in access to food within the family, overwork and inadequate food. The energy output of women is not compensated by proportionate intake of food, the ratio of intra-household male-female food distribution was 2:1. It is estimated that more than half the world's households cook daily with unprocessed solid fuels and have grave health implications. These toxic pollutants released in biomass cooking pose great health risks to women and female child.

Health

Nutritional deficiencies have direct health consequences. In addition, because of the nature of tasks proper, women perform fetching drinking water, washing clothes in streams or transplanting rice, they are more directly exposed than are men to water-borne diseases, and to the pollution of rivers and ponds with fertilizer and pesticide run-offs. In some regions, several times more of the acceptable levels of chemical residues have been found in the milk of nursing-women agricultural workers. The burden of family ill health associated with water pollution likewise falls largely on women.

The reduction of water consumption, because of the time and labour involved in collection, leads to many geneto-urinary and reproductive tract infections affecting personal hygiene of women. The health costs of the nexus between energy scarcity, the resultant dependence on biomass fuels, the exploitation of human energy to meet basic needs and gender division of labour are extensive, effecting widespread protein malnutrition and poor health among women and girls, chronic anemia, high material morbidity and mortality, poor reproductive outcomes, reduced chances of infant survival, depletion of women's health from overwork, inadequate food and repeated child bearing. This burden is carried by millions of poor women and children who are already the most socio-economically disadvantaged segment in most countries of South Asia.

Social Implications

Population displacements arising from the submersion of villages due to large dams, or from large-scale deforestation, disrupts the social support networks with kin and other villagers, built up especially by women. These networks provide small loans of food and cash, or labour exchange and tide poor families through periods of shortage. Their disruption usually goes uncounted in cost benefit exercise of large irrigation schemes.

On Time

Since it is women and female children who mainly collect firewood fodder it is their time and energy which gets extended when forests and VCs decline. In many areas distances travelled and time taken for these tasks has increased manifold. There is very low return for each unit of human labour and time invested in vital subsistence and productive activities for instance a round trek of 6 to 7 kilometres, may yield firewood only for one day's cooking. In contrast a middle class urban household spends less than one tenth of time and labour for the same result. Meeting basic needs for survival consumes enormous quantities of their time and labour, which cannot be diverted to more productive or life enhancing activity.

On Women's Indigenous Knowledge

Food and medicinal items from forests demand an elaborate knowledge of the plant species, which these women have gained through long experience. Also, in many regions, peasant women responsible for indigenous seed selection are poor. With the large-scale shift to hybrid seed, control has passed into the hands of national/international laboratories. While traditional systems are being devalued, the women who depend on this knowledge have little institutions, which create what is seen as knowledge and modern technology. In this sense, what we see today is not sustenance for the poor and for women; it is of knowledge critical for generating sustainable livelihood systems.

While identical pattern of gender effects are reflected in most parts of India and South Asia the intensity varies due to gender vulnerability, environmental risk and incidence of poverty in a region. Gender vulnerability is especially high in the northern states characterized by highly female adverse ratios, low female literacy, strong female seclusion ideology and women's limited effective access to property, especially, arable lands. South India in contrast has better sex ratio, more female, literacy, absence of female seclusion practices, and greater effective rights in property (Kerala).

Similarly, environmental vulnerability, as measured by forest cover and rainfall levels, is much higher in north, west and western South Asia than in north-eastern belt. Poverty regions are highest in eastern India. Women are best where vulnerability on all three counts is low.

Gender, Environment and Poverty

While the need for gender justice is universally accepted and eradicating discrimination both formally and substantively is a prerequisite for creating an egalitarian society, substantive justice cannot be achieved as long as basic needs of women remain unmet and they continue to be deprived of equal access to productive resources.

Environmental Refugees: Women in Urban Slums, Squatter Settlements and Industrial Zones

"By the turn of century, the majority of the world's population will be living in cities. While urban settlements particularly in South Asia countries, are showing many of the symptoms of the global environment and development crisis, they nevertheless generate 60 per cent of gross national product and, if properly managed, can develop the capacity to sustain their productivity, improve the living condition of their residents and manage natural resources in a sustainable way."[20]

These cities are the main hub of the economic activity but already under great stress due to overstretched and inadequate infrastructure. Housing is in a critical short supply. This migration has been basically from rural, hilly, deserted and other degraded areas. The 'Shramshakti' Report (1988) noted with concern the failure of the 'trickle-down' theory of development and perceptively commented as follows:

The limited achievements in the anti-poverty programmes have been more than offset by the problems of land alienation and environmental degradation, increasing agricultural poverty, concentration of resources in a few hands, increasing polarization of the rich and the poor and polarization between urban and rural areas. Women have been particularly hard hit by these developments.

In spite of their important role in agricultural production, the land reform measures like land ceiling and distribution and tenancy reforms have not benefitted them because land has rarely been in the name of women. More and more common property resources have been taken over by the government or have been privatized. This has added to the burden of women who are almost solely responsible for collecting and fetching water, gathering firewood, small game and other forest produce. Environmental degradation like deforestation and commercialization of forest resources, indiscriminate tapping of groundwater resources have further aggravated women's problems. Increasing agricultural poverty has led many men to migrate in search of work, leaving their families behind to face the consequences. To this could be added large-scale displacement by dams and industrial projects, which do not always bring benefits to the local population. The displaced persons become pauperized and their women are worst affected since rehabilitation plans make no attempt whatsoever to help the women to gainful employment to attempt to compensate the loss of the economic activity in their previous locations.[21]

And where do all the environmental refugees find shelter? In 1981, out of a total urban population of nearly 160 million, 32 to 40 million, i.e., 20 to 40 per cent were in slums alone. While population has grown the infrastructure has not, thus leading to increase in slum dwelling women.

In this steady stream of those coming to the big cities the men far outnumber women, still more and more women are beginning to migrate to the cities, sometimes with their families and sometimes even alone. Many of them end in the red-light areas—the dens of prostitution—a large number become domestic servants to the relatively affluent middle-classes living in proper houses, women construction workers, young and old, with head-loads, taking them up many flights of steps can be seen any day. Illegal factory manufacturers, not obeying any labour law at all, accommodate a large number of women, who are paid on an average of two-thirds or three-fourths of what men would be paid. The underpayment of this 'underclass' ensures that they will never be able to extract themselves from illegal squatter settlements of slums. In a sense, the economic and social strait jackets into which they find themselves squeezed, perpetuate their poverty by 'institutionalizing' it. They occupy the least desirable spaces. Often the very location of their settlement is a threat to their lives.

"As in the forest and rural areas, the provision of food and fuel becomes women's main responsibility. The bad coal ash, mixed with mud and turned into fuel brackets is usually their main fuel. Cooking, cleaning, living without any sanitary facilities, they become a prey to many water-borne diseases—cholera, typhoid, diarrhea, particularly among children, malaria and intestinal worms claim many adult childrens' lives. Women and girls are the worst sufferers because they have under-nourished physique and relative paucity of medical facilities, provided to them by their own families, takes its toll. Women because of this adverse effect of environment on their health die more in childbirth; pregnancy doubles the risks of health from common diseases such as pneumonia and influenza."[22]

"Calcutta has been known as the cholera capital of South and South-East Asia, Kanpur for long has been known as the tuberculosis capital and now Delhi is getting the dubious distinction of the world's asthma capital. As the water levels in the rivers, Ganges, Yamuna or Gomti, the main source of drinking water goes down, the incidence of epidemic jaundice and deaths become an annual feature along with epidemics of cholera, euphemistically called gastroenteritis, and who is the greatest sufferer; the undernourished women and the girl child, who would not even be taken to the hospital, just left there to die or survive".

The urban environmental issues are complex and can be listed as water, sanitation, solid wastes, industrial pollution, transport, urban household energy, land planning and development including built-up environment. These issues are cross-sectoral and get further complicated because of multiple jurisdictions. Apart from the general environment, the home environment is again seriously inadequate. Insufficient and erratic piped water supply, difficulties of cooking fuel, and the requirements of privacy in bathing and sanitation pose special problems for women, 70 per cent have no access to community latrines, lighting or drinking water. Projects to identify women's requirements of science and technology inputs for improving the quality of life of women in slums and squatter settlements all point out to sanitation as their main difficulty. Grassroots workers are faced with demands for 'sulabh shauchalayas'. Cooking on inefficient stoves and bad quality fuel makes women more prone to respiratory diseases. It is like smoking 200 cigarettes a day. Delhi health statistics are ample proof of this.

The macro-economic environment, as noted earlier, has reduced the share of agriculture and increased that of the manufacturing and service sectors, in the Indian economy and its GDP. Women's cultivator status has gone down and marginalization has vastly increased. The 'Shramshakti' Report found that 93 per cent of women are engaged in the unorganized industrial sector. The incidence of self-employment amongst women marginal workers (subsidiary status) is far higher nearly 82.2 per cent in urban areas. Moreover, the glaring gender disparities in earnings compound their problems, and increase with the age group of workers. Traders, middlemen, contractors and big companies in the corporate sector also exploit them as piece-rate workers.

Developmental Model, Environmental Conservation and Women

There can be no denying the fact that the continuation of this growth model will lead to further ecological destruction, to more inequality and to more poverty. This is bound to affect the women and children the most. Consumption patterns are leading to growing sexual abuse and direct violence against women and children, both in public and within the family. Sex tourism to countries of the South, and even the growing perversity of sexual abuse of little girls are symptomatic of this diseased mentality, and commodification of every human value, any kind of dignity for the poor women and the girl child. But for the corporate financiers these are only service industries, founded on the indignity of women and the girl child; all ancient civilizations are being trod under foot with a goodbye to all values of decent human life for the poor, who have no option but to sell themselves, resourceless and powerless as they are.

The sectoral and reductionist approach to development fails to see the interconnection between forests, livestock and crops, and also the crucial role played by women in ecological conservation. Vast scale irrigation and act of chemical fertilizers also has led to salinity of land, pollution of underground watertable and creation of wastelands. Women's roles on social forestry need to be realized better, regeneration of wastelands and substitution of micro-biological nutrients for crops, which again would need the kind of care that women alone can give. All these allied activities which are ecologically and economically critical reveal agriculture as women's major occupation, and still due to degradation of agricultural land and its alienation for other purposes, and cash crops has reduced women's cultivator status and increased casualization of their labour in farm and allied activities.

Women's traditional knowledge of conservation strategies needs to be supported by a vast network of extension and training and change in research priorities in agriculture to remove this gender blindness of modern agricultural education, based on male-dominated western patterns of agriculture. A vast potential of employment lies in involving women with wasteland management through social forestry programmes.

The unequal distribution/ownership of resources within and outside the household that environmental destruction exacerbates women's problems in a way very different from that of men. Environment conservation and poverty removal thus become two faces of the same coin—an unavoidable challenge.

Resistance and Awareness

The noted negative effects of gender inequality and environmental degradation, however, have not gone unchallenged by those affected. As early as the eighteenth century in the year 1787, Amrita Devi gave up her life with 363 women against the king of Jodhpur to save the Khejri trees in Rajasthan by hugging and embracing trees around with arms. In the 1970s, Gaura Devi and a group of women hugged trees in Tehri Garhwahl to protect their felling by forest departments. Women like Sharmishta Jagawal, are leading tribal women's ecodevelopment movement in Gujarat. Environmental activist Medha Patkar has been fighting for the rights of tribal, and forest submergence in the Narmada Sagar dam. In Andhra Pradesh, Vasantha Kanibers leads an organization of villagers to reclamation of degraded lands through traditional farming systems and organic compounds. Awareness to government policies on potential loss of biodiversity is being lead under Navdanya—a seed movement of Vandana Shiva.

The last two decades have seen the emergence, both of women's groups and environmental groups, the former protesting the gender bias in existing patterns of development, the latter their high environmental costs. In some cases, gender and environmental challenges have overlapped. Certainly, women have been significant actors in major environmental movements. As such women's involvement in such movements, needs to be contextualised. These movements have emerged mainly in hill or tribal communities among which women's role in agricultural production has always been visibly substantial and often primary. Rather, poor peasant and tribal women's responses to environmental degradation can be located in their everyday material reality; in their dependence on natural resources for survival and the knowledge of nature gained in that process. From the plethora of examples cited above one can very well conceptualize why ecological destruction and planetary health are feminist issues.

This brings us to the ongoing debate on how we should conceptualize the relationship between women and nature. In patriarchal thought, women are identified as being closer to nature and men as closer to culture. Nature is seen as inferior to culture; hence, women are seen as inferior to men. Because the domination of women and of nature has occurred together, women have a particular state in ending the domination of nature. The feminist movement and the environment movement both stand for egalitarian systems, and need to work together to evolve a common perspective, theory and practice. In particular, ecofeminism calls upon women and men to reconceptualize their relationships to one another and to the non-human world, in non-hierarchical ways. South Asian variants of ecofeminism share many of these assumptions.

Environmental Conservation and Women Empowerment

The women's perception in South Asia recognizes that meeting human and social needs is a goal of higher importance than merely economic progress for its own sake; such perception must be integrated with problems of environmental degradation and poverty arising from inequitable access to necessary survival resources. Women must be empowered at all levels to contribute to the achievement of principles of equity, social justice and ethics, as they are the working hands and units of implementation. It is now an undisputable fact that increasing environmental movements and women's involvement made significant inroads and impacts in eco-development the world over and labour-intensive South Asia in promotion. Nature-dependent livelihoods, demands for environmentally sustainable policies and equalitarian access to natural call for underpinning importance of greater participative approach in eco-conservation and programme implementation.

It is imperative, therefore, that South Asian states should recognize women's role as managers and conservators of natural resources and should involve women in the decision-making process as equal partners in agriculture, forestry, resource mobilization, marketing and sustainable development. There is also a need for participatory movement of women in eco-development programmes. If both men and women are involved in decision-making with accommodation and articulation to incorporate respective priorities, they may be able to devise a management system equally responsive to their respective priorities through negotiation, e.g., Mahuas Mango. What is generally seen in community forest management by state or individuals is that women continue to lack in giving out explicitly their views in community decision-making because of the unquestioned assumption that their household men will automatically take care of their interests. Due to the ignorance, subtleness and invisibility of the processes at work, the affected women are not even aware of the happenings and implications. Women have already lost out in large land distribution programmes undertaken in politically enlightened democracies like India because land titles have been given to men as assumed "heads of households". It is, therefore, crucial to see that women are not left out again in the process of managing country's resources. The issue is not only of gender equity but also of sustainability, justice approach and involvement.

It is also critical to integrate recognition of women's identity as well as different needs, priorities, and aspects in decision-making and analysis at the community level. Admittedly, women should participate in equal terms in all decision-making for development of eco-friendly sustainable regimes helping in emergence of economies based on environmental emancipation in South Asia.

It is also imperative that women in positions of power and influence should not lose women's perspective on development and environmental issues and prompt men to reconcile to this need for common good.

States should ensure that women participate and benefit from rural development on the basis of gender equality and have access to agricultural credit and loans, marketing facilities, appropriate technology, equal treatment in land and agrarian reforms as well as in land resettlement schemes. They must also be assured of land use and ownership, education and health services.

Due importance should be given to the traditional knowledge systems of women and they should be educated and given extensive training on modern agricultural methods and technology.

Basic provisions of Agenda 21 regarding women's rights should be implemented.

Women approach of rearing and nursing instinctively overweighs that of men. This could be harnessed more effectively for conservation of environment and its development by increased education and awareness programmes.

Women by acting together, can help change policy, protect the environment, improve standards of living and dispel current economic disparities if given opportunity and supportive hand of man. Although involvement of women in decision-making is still limited, it is likely to grow once there is a greater realization that involving women in natural resource protection and regeneration is crucial not only for women's empowerment but also for success and sustenance of the environment it self. Indeed, ecofeminist theory is based on community—based on knowledge and valuing, and the strength of this knowledge is dependent on the inclusion, flexibility and reflexivity of the community in which it is generated. Without the cooperation of women as equal partners, rules instituted for protection and conservation will not work effectively, given their primary role in collection and gathering of survival needs.

Notes

1. Pietila, H., "Alternative Development with Women in the North" in *Alternatives Akademilitteratur*, Johan Galtung and Maris Fribergs (eds.), Stockholm, 1986: 26.

2. Gaard, G. and L. Gruen, "Ecofeminism: Toward Global Justice and Planetary Health" in *Society and Nature*, Vol.2, No.1, 1993:2.

3. *Forward Looking Strategies for the Year 2000*, (1985), U.N. Publication, New York.

4. *India's Country Report: Women and Environment*, presented at the 4th World Conference on Women, Beijing, 1995: 2.

5. Antrobus, P. and Bizot J., "Women's Perspectives: Towards an ethical, equitable, just and sustainable livelihood in the 21st century", in *Women and Sustainable Development*, Women in Action Journal of ISIS International Issue, March 1992 and January 1993, Quezon city Maine: 35.

6. Shiva, V., *Staying Alive: Women, Ecology and Survival in India*, Kali for Women, New Delhi, 1989:xvii

7. Akeroyd, V.A., "Gender, Food Production and Property Rights: Constraints on Women Farmers in Southern Africa" in *Women, Development and Survival in the Third World*, London: Longman Group, 1991: 139,17.

8. Shiva, V., "Where has all the Water Gone?: Women and the Water Crisis" *Ecoforum*, Vol.10, No.3 April 1985:16.

9. Waring, M., *If Women Counted: A New Feminist Economics*, San Francisco: Harper Collins, 1988.

10. Brown, L.R. et al., *State of the World: 1992*, New York: W.W. Norton and Co., 1992: 4.

11. *Eighth Five-Year Plan*, Vol.II, Planning Commission, New Delhi, 1992: 94.

12. Jodha, N.S., "Common Property and the Rural Poor", *Economic and Political Weekly*, 1986.

13. "Environment and Development, Traditions, Concerns and Efforts in India," National Report to UNCED, June 1992, Ministry of Environment and Forests, Government of India: 8-9.

14. Jodha, N.S., op. cit., 1986.

15. Shiva, V., *The Violence of the Green Revolution*, London: Zed Books Ltd., 1991: 16.

16. Agarwal, B., "Ecofeminism" in Hindu Survey of Environment, Chennai, 1995.

17. Shiva, V., "Most Farmers in India are Women" FAO, New Delhi, 1991: 4

18. Ibid., 5.

19. Ibid., 13.

20. Agenda 21:58.

21. 'Shramshakti' Report of the National Commission on Self-employed Women and the Women in the Informal Sector, Government of India, New Delhi, 1988: 7-8.

22. Swarup, H. and Rajput, P., "Development Policies and Environmental Measures: A Case Study of India as it Affects Women", in *Environment and the Disinherited Half: Some Third World Perspectives*, Manushki Publications, Kanpur, 1992: 31.

25

Environmentalism, Development and Human Security: An Obvious Connection

Ashutosh Saxena

The original content of philosophy of fundamental freedom was limited to civil and political rights of individual, often referred to as first generation rights, which at large call for a negative obligation on governments to desist from interfering with the exercise of individual liberties. The expansive nature of human rights clubbed with the realization that without guaranteeing economic, social and cultural rights, full enjoyment of civil and political rights was not possible led to the inclusion of economic, social and cultural rights, popularly described as second generation rights. These second generation rights place a more positive duty on governments to act in order to ensure the realization of these rights. The concept of human rights in its comprehension, has now travelled to encompass what are called third generation rights, which include: right to self determination and security, right regarding as belonging to people rather than individuals, right to sovereignty over natural wealth and resources of the country and right to development as well as right of disadvantaged groups to special protection.

Emergence of new and very powerful movements like feminist movement, movement for environmental protection and sustainable development, movement for protection of minorities and indigenous people and their culture, movement for abolishing of child labour and all forms of exploitation contributed to taking recourse to human rights and its language more frequently. In fact, these movements have contributed in great measure to the increasing of human right concepts in national contexts. Respect and realization of human rights requires evolving a culture that is more sensitive to the basic needs of every human being. It respects the need for ensuring to everyone, justice—social, economic and political and provides fair and equal opportunities for growth and development to every individual and group of people.

Often efforts for freedom from want, leading to economic and structural growth also fastens the pace of environmental degradation. And so, security for one dimension of human well-being becomes a threat to another dimension. As a result, mankind today is facing a serious threat of existence and survival due to the continuous depletion of environment in the name of development. Climatic changes, global warming, greenhouse effect, depletion of ozone layers, increasing frequency of earthquakes, droughts, floods, epidemics, increasing of new types of diseases, etc., are all results of environmental pollution and degradation.

Thus, the core demand of human right, today, is the right to human survival and existence. Human survival and existence include not only decent and safe human life but also protected socio-economic, political and natural environment, as such an environment is the precondition for obtaining sustainable development. Here, a healthy natural environment has been taken as a basic factor determining human security.

The present paper attempts to discuss this key issue of human security in the context of development vis-à-vis environment. The paper, taking Indian philosophy and literature as reference study for developing nations, attempts to search for viable strategies for achieving sustainable development.

It is a great irony that many of the challenges we face today are the unintended consequences of our efforts to enhance the security and welfare of humankind. Unfortunately, our efforts have involved:

1. extracting resources (such as fish, fresh water and timber) faster than they can be replenished;
2. loading toxic and other waste materials into our land, water and air faster than they can be broken down and neutralized; and
3. drastically modifying large ecosystems (from rain forests to coral reefs) such that they can no longer support many species or effectively provide important environmental services such as climate control.

As a result, throughout the world, humankind is experiencing scarcity (especially of food and water), microbial invasion, loss of option values, and excessive exposure to toxic substances and particulates. It is not, however, experiencing these things in a uniform manner. The careful manipulation of economic, political and cultural systems allows some individuals and groups to direct many of the adverse consequences of environmental change towards weaker entities or future generations.

Concepts Involved

Environmentalism, human development and human security are the three key concepts used in the present paper. Concepts are used in the broader framework of human rights. These concepts are contextual and are explained below for understanding purpose:

Environmentalism

Environment is a network of physical, chemical and biological elements. All these elements regulate and control the environment through mutual interactions and reactions. Human beings are an integral part of environment; they are products of environmental process and exist only in coordination with environment. Any imbalance between them lead to disastrous results.

Man-environment interrelationship has always been a fascinating aspect of study. Environmentalism is one such ideology, which focuses upon the concern about environment and efforts to save environmental degradation. The ideology of environmentalism believes that nature is supreme and the life process is controlled by nature. Thus, the fundamental belief of environmentalism is that the behavioural pattern of human beings is determined by nature. Ancient Greek philosophers like Hippocrates, Aristotle, Stray, Bodein, Montesque, German philosopher Hemvolt and Charles Darwin, are among such prominent philosophers and scholars, who have defined man-environment relationship and the dependence of man on nature.

A second line of ideology, which includes Feber, Blash, Brunce, Karl Sawer, etc., does not believe in this eco-centric approach to environment. They rather believe in the ideology of 'possibilism', which advocates man-nature relationship as mutually influencing relationship. Man through his power, capacity, mental ability and technological advancement has always exploited the nature to his benefit. Possibilism believes in the supremacy of man over nature. Man has the capacity to alter the nature according to his desire. Both environmentalism and possibilism are extreme approaches, one keeps nature in the centre, and the other keeps man in the centre. A third approach, which in fact is a midway, is known as neo-determinism. It believes in equi-importance to both man and nature. Grift Taylor is its main exponent. According to him, both man and nature influence and affect each other. Ecological approach to environment, close to neo-determinism, believes that too much interference of man in nature ultimately harms the man because equilibrium is destroyed. Too much human interference of man in nature for the sake of development leads to destruction.

Human Development

It has been recognized the world over, and also substantiated by our own experience that democracy, development, cultural pluralism and preservation of human rights are interdependent. Democracy is not a mere form of government, it is a set of values and a way of life that ensures respect for identity, choices, capacities and abilities of every human member of the society. Similarly, development is not merely development of material resources but also of people, infrastructure and betterment of quality of life of every member of society. Thus, respect of every member of society as a distinct human being, and, welfare of every human member, is the common core of democracy as well as development. Society that lacks or fails in its commitment to the preservation of human rights can hardly boast of democracy or development. In such a society, benefits of development are appropriated by a privileged few and respect and recognition of its members is largely determined by their possession and status.

No meaningful development is possible without processes of development being accountable to the real holder of power, namely, the common masses. Most of the activities undertaken for the sake of development result in mass uprooting of indigenous people, degradation of environment, unbalancing ecology due to an imbalanced approach to development and result in gross violation of human rights of those affected. It is an accepted fact today that sustainable development has also to accommodate preservation of environment and ecological balance. Without such a balanced development a stake in democracy can hardly take roots in the minds of masses. Ironically, the modern imbalanced approach to development is short termed and self oriented, which ultimately will lead to destruction of mankind itself. Thus, the most critical threat, today, is not only the threat to social-economic, political, military and cultural security but also security and preservation of environment and nature.

Human Security

The concept of human security implies freedom or protection from threats to human well-being, be it military, economic, social and resource (food or otherwise). In its comprehensive sense human security demands rights, which could no more connote merely to the physical well-being of an individual, but must expand to comprehend all those conditions in a society that makes human existence possible with dignity and honour. The industrial revolution has threatened human security in the name of development. The environmental depletion due to industrial growth has put man in danger.

All the systems in a social framework have their own mechanism of human security. But certain elements may appear in the system, which are functioning negatively to human security.

In general, the following can be identified as important elements of human security:

Table 1
Elements of Human Security

System-Promoting Mechanism	Security-Promoting Mechanism	Insecurity-Mechanism
Economic	Wealth	Poverty
	Welfare Policies	Inequity
Political	Law	Corruption
	Legitimate Force	Unlawful Use of Force
Cultural	Social Identity	Discrimination
	Justice	Injustice
Demographic	Low Birth Rate	High Birth Rate
	Urbanization	Rapid Population Flows
Ecological	Life Support	Scarcity
	Raw materials	Disease
		Environmental Change

The above-mentioned elements are not operating in isolation in a society. Rather, they produce a network of relationship, where they influence each other. The network of this relationship can be seen in Chart 1.

From a human security perspective, environmental change is important in two ways:

First, it can itself be a source of insecurity. Even if the state does not register a given form of environmental change as a threat to its core

Chart 1
Human Security Network

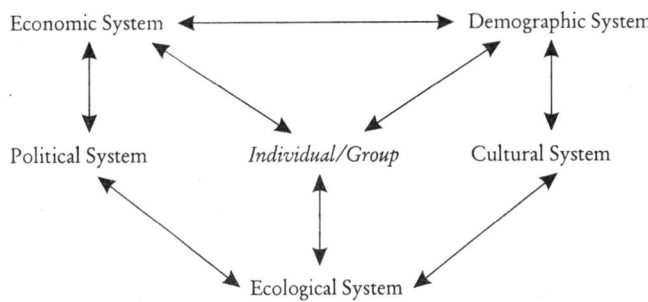

values or national interests, fragments of its citizenry may feel otherwise. On the other hand, if the state does appreciate the security relevance of environmental change, it may still be constrained in terms of its response insofar as the problem has diacritical transnational dimensions.

Second, modes of environmental change can exacerbate other real or potential forms of insecurity such as poverty, discrimination, desease, agricultural loss, potentiality losses, etc.

Insofar as these two general claims are valid, it is crucial that analysts and policy makers consider security/insecurity from the comprehensive vantage provided by the analytical framework of human security. This will foster a better understanding of causal relationships, which are almost always network-based. Better understanding will in turn help reduce the likelihood of recommending or implementing bad policies. For example, an attempt to reduce insecurity by alleviating poverty may fail if it involves increasing environmental degradation or scarcity, creating a situation in which short-term gains are almost certain to be undermined by medium or long-term losses. There is indeed much truth to the environmentalist dictum: "everything is connected to everything".

Human Security, Development and Environment: An Obvious Link

The above issues lead us to formulate the hypothesis that development, security and environment are interlinked. Approach to sustainable development must involve methods and ways to bring development without destroying environment. The Rio Summit (1992) had officially linked environment and development issues, including an explicit recognition that poverty itself is a driving force behind a large share of environmental degradation. This connection between environment, development and human security can well be understood through following cybernetic multiloop simulation model:

Model 1

The above model shows that development has twofold effects on human security. On the one hand, it is positively related to human security, implying that modern type of development may provide materialistic gains and freedom from threats to food and discomforts of life. But this is a short-term approach. The same development strategies due their inherent tendency of destroying the environment, adversely Effect human security and put human beings to more dangerous threat of survival and existence itself. The consequences may not look alarming in the beginning, but will lead to devastation in the longer run. Environment will keep on depleting and soon ecological imbalance will threaten the human security. The above projected model leads to two pertinent questions:

1.   If we accept the linkages in the way they have been shown in the model, what would be alternative strategies, particularly for developing countries like India to achieve sustainable development in the realm of cost benefit framework.
2.   Should we search the alternative strategies in the history and philosophy of Indian environmentalism or in the present approach of western environmentalism?

Intervening Factors

The process of scientific and industrial development, which started in western countries long back, has a long history of destroying environment for economic, scientific and industrial growth. Today, these countries directly and through their controlled agencies are showing much concern over environmental issues largely shifting the responsibilities of environmental hazards on developing countries. Today, all the countries of the world have recognized the fact that environment is a shared heritage. It is the responsibility of world community as whole to protect it. The world attention to this fact was drawn in Stockholm World Environment Conference in June 1972, where 119 countries accepted principle of 'one earth'. The conference declaration included:

1. Human beings have a fundamental right of living in free, equal and congenial environment.
2. Environmental pollution has emerged as a continuous danger to human survival.

Thus, it is the responsibility of each individual to protect the environment from pollution. But, unfortunately, the big industrial nations did not pay any heed to this direction. Second summit, known as Earth Summit, was held at Rio de Jenerio in 1992, where 170 countries participated and an Agenda 21 was proposed to conserve and protect the environment. It is important to mention here that rich and developed countries like USA contribute to a large extent to environmental degradation through their addiction to automation. In fact, developed countries like USA have a long history of destroying environment for economic, industrial and scientific growth. Now these countries directly and through their controlled agencies like UNO, WTO, IMF, etc., are showing much concern on environmental issues, largely shifting the responsibilities of environmental hazards on developing countries. According to UN report (2000-2001), the western rich population of the world is rapidly consuming the resources of the earth. It is an astonishing fact that the amount of environmental pollution created by one child born in western industrial country is equal to 50 children born in a developing country.

In August 2002, another World Environment Conference was organized at Johannesburg (South Africa) after ten years of Earth Summit. There was a clear and deep divide between rich and poor countries over environment issues. The western countries like USA were reluctant in giving environment protective economic package to developing countries. These countries were keener in implementing their liberalized economic packages to enhance their own economic interests. It is high time to smell politics behind environmental propagation by developed countries, as these countries do not intend to allow developing countries to grow industrially, scientifically or otherwise in the name of environment through bindings of WTO, IMF, etc. Through implementation of their policy of LPG (liberalization, privatization and globalization), these countries are setting up their industrial units in developing countries causing environmental hazards to these countries and also with the intention of economic extraction. The cost of treatment plants and other mechanisms to control pollution is so high that running of those units becomes economically unviable, particularly for small and middle entrepreneurs in developing countries.

There is no denial to the fact that most of the developing countries including India are facing challenges of environmental pollution and degradation, resulting in constant threat to human security. Various political, social, economic, policy, planning and attitudinal factors can be identified as responsible for this decay, but here our primary concern is to search for viable alternative solutions to the problem because for all human beings the right to decent and safe life is one of the basic human rights. Under the world political scenario, the tendency of dependence over rich and developed countries for countering the environmental challenges must be shifted to developing indigenous models, which may suggest methods to healthy environment and dignified human survival.

Sustainable Development in the Framework of Indian Environmentalism

Respect for human rights as a part of its social philosophy has existed in the Indian ethos for a long time despite its aberration intermittently. During this same period, a very different perspective—the ecological worldview—has also gained clarity and become influential in many parts of the world. It contends that humans are recklessly transforming and destroying nature on a grand scale. Its frontline objective is to secure the environment from the unprecedented threats posed by uncontrolled human activity. By following nature, that is, by adapting ourselves to natural patterns, rhythms and thresholds, many of its proponents suggest, we will not only cease those activities that are destroying our life support system, but we may also recover some of the rich purpose of life that has been lost in our consumer societies—spirituality, beauty, truth and simplicity.

The concern for environment has always been the focal point of ideas of philosophers and scholars in India. Today, there is great hue and cry over environment issues, particularly since the western concern over environment and ecological pollution has gained momentum. India is also affected by it. If we look historically, the problem has intensified in last 20 years, largely due to modern approach to development.

The ancient Indian literature has always taken care of nature and environment. Ravindra Nath Tagore had once said that Indian culture is not the culture of villages or city, in fact Indian culture is the culture of nature and forest. Our historical traditions, philosophy and culture have been originated only in nature and forest. Modern approach to development has encouraged consumerist culture. This consumerist culture has turned into an enemy of nature. Man has tried to be the master of nature and he has spread the technological network, which exploit the nature. According to Indian standpoint, man is son of soil—not master of nature (Chatak, 1992). During the ancient time, the problem of ecological imbalance was non-existent, but still Indian saints, philosophers and scholars were always conscious towards environmental issues is quite evident in Indian literature. During the Vedic period, man was considered as part of nature. The genesis of universe itself imbibes the coordinative relationship between man and environment. Nature is the product of *yajna-purush*. Man-nature is made up of earth, air, water, space and sun. It is important to note here that man has nowhere been given priority or over-rights in this man-nature relationship. Man is like any other element of nature. This is one of the reasons why man prays for welfare of all flora and fauna along with their own welfare. Even for religious purposes, trees were not uprooted; only leaves and branches were used. Kautilya in his *Arthshastra*, has emphasized on the conservation of various elements of nature. He has suggested that one of the prime duties of king is to protect forests, animals and other natural objects. *Arthshastra* contains detailed description and analysis of agriculture, weather phenomena, floods, droughts, etc., and their remedies. Kautilya has also suggested that those who play with nature must be punished. *Mahabharata*, another great Indian epic, is equally sensitive to nature. The epic focuses upon the pure and healthy environment. It contains that man and other natural beings must coexist peacefully.

Mahabharata considers earth as a comprehensive whole and all visible and non-visible objects are part of it. It is the responsibility of man to systematize his life pattern, to contain his desires. It is a must for a good physical environment. Healthy environment is possible only through mutual coordination among the human beings. *Manusmriti* is also conscious of ecological balance. According to Manu, man is one of the lacs of creations by God. It is the responsibility of man to protect living and non-living elements of nature. Human existence is possible as ecological balance is maintained.

In order to achieve sustainable development in Indian environmental framework, following approaches, based on history and philosophy of Indian environmentalism, can be suggested:

Qualitative-Normative Approach

Respect for nature has been an eternal value and culture of Indian tradition. The quantitative development believes in more GDP, increasing per capita income, more materialistic comforts in life, whereas qualitative approach focuses on establishing equilibrium between man and nature. Man is part of nature and love and preservation of nature are the natural elements, which must be upheld. There is a need to redefine the concept of development in the value framework. The approach suggests that focus should be on developing the human beings in a qualitative manner and not on enhancing the consumerist culture in quantitative manner. The essence of Indian philosophy is that man is always more important than matter. So efforts should be on cultivating the inner values and culture in human beings.

Containment Approach

Industrial economy survives on creating more desires and lust among people. Soon, desires take over the value structure and an unending process of demand and supply begins. In the process, the social structure and natural fabric of the environment are severely affected. This modern life style has transformed the individual personality. Materialism has given way to consumerism. As a result, the society and environment both have detoriated. The Indian environmentalism and approach to sustainable development suggest an attitude of self-containment, i.e., a sense of *atma santushti*. Man should have control over desires and lust. Simple living and high thinking should be the motto in life. The consumerist culture takes them against the nature.

Nature-Oriented Approach

In all the Indian literature, mythology and philosophy, which include *Vedas, Ramayana, Mahabharata*, etc., nature has been given priority over everything. In fact, Max Scheler traced to the cosmic love of nature as a basic value in the Indian tradition. C.C. Jung and many others belonging to the traditional school of thought have also held similar views. Human beings are supposed to be made up of five natural elements—fire, water, sky, air and earth. Nature has been considered as the first form of worship. The first known objects of worshipping were natural objects. The nature-oriented approach has great faith and respect towards nature. It believes that if you respect it, nature will respect you. The western development process has expedited the exploitation of natural resources. Most of these resources are non-renewable. Excess production increases the level of pollution on the one hand and diminishes the natural resources at the same time. The approach of 'more production – more development' has created crisis of survival of civilization before us. The nature-oriented approach suggests extracting only that much from nature, which could be given back to nature. *Gita* propagates the principle of mutual and equal participation of man and nature. This is *yajna*. One of the basic purposes of performing *yajna* is to maintain the cycle of natural phenomena. The 3/12 verse of *Gita* contains that if a person does not fulfil the responsibility of returning of what has been given to him is immoral. Nature gives us fresh water, air and food, so it is our responsibility to maintain it. Lord Krishna believed that controlling the individual desires would automatically regulate the proper cyclic circulation of nature. No civilization can grow on the premises of exploitation of nature. All forms of *yajnas*, worshipping the grains and even celebration of religious and social festivals are linked to nature and environment. So, there is a strong need to develop ethnocentric attitude towards environmentalism.

Gandhian Approach to Economic Development

Out of many western and indigenous models prescribed for sustainable economic development, the Gandhian model seems to be most befitting for developing countries like India under the present global socio-economic scenario. Critiques may have their own arguments against the model in the light of LPG, but under the Indian social framework, it is to be decided, "do we want vertical growth or horizontal growth?" Vertical growth does not care for anything except profit, thus it not only runs over the human beings, but also over its originator—nature. On the other hand, horizontal growth takes everyone along. Gandhian approach believes in horizontal growth model. The approach stands for technological-industrial advancement and scientific temper, but as long as it does not intervene with the social and natural fabric of society. Economic growth must be in alignment with nature. Economy should be self-sufficient and indigenous. The Nobel laureate Prof. Amartya Sen also subscribes to same view.

The above approaches combined together will definitely inculcate environmental ethics amongst the human beings, which will accelerate the pace of development in correct and positive direction, so that the primary human right of human security is protected.

References

Augustine, John, *Strategies for Third World Development*, Sage Publications, New Delhi, 1989.

Bitchel, B. Robert, *Environment and Behaviour*, Sage Publications, New Delhi, 1997.

Buell, John and DeLuca, Thomas, *Sustainable Democracy*, Sage Publications, New Delhi, 1996.

Chatak, Govind, *Paryavaran aur Sanskriti*, Takshila Publications, New Delhi, 1992.

Connely, James, *Politics and Environment*, Taylor & Friends, London, 2002.

Feber, *Crisis Reading: Environmental Issues*, McMillan Publications, London, 1971.

Frost, Mervyn, *Constituting Human Rights*, Taylor & Friends, London, 2002.

Garg, Bansal, Tiwana, *Environmental Pollution and Protection*, Deep and Deep Publications, New Delhi. 1997.

Glan, L.A., *Ecological Crisis-Reading for Survival*, Harcourt Brace, New York, 1970.

Gutenberg, Tom and David Mickie, *Eco-impacts and the Greening of Post Modernity*, Sage Publications, New Delhi, 1996.

Hempel, C. Lamont, *Environmental Governance: The Global Challenge*, East-West, 2001.

Kaushik, Asha, *Globalization, Democracy and Culture: Situation of Gandhian Perspective*, Avishkar Publications, 2002.

Koithra, Verghese, *Society, State and Security*, Sage Publications, New Delhi, 1999.

Kumar, H. D., *Sustainable Human Ecology*, East-West, 2001.

Macnaghten, Phil and John Urry, *Contested Natures*, Sage Publications, New Delhi, 1998.

Mehta, Shekhar, *Controlling Pollution*, Sage Publications, New Delhi, 1997.

Pinto, Vivek, *Gandhi's Vision and Values*, Sage Publications, New Delhi, 1998.

Scott, Lash et al., *Risk Environment and Modernity*, Sage Publications, New Delhi, 1995.

Sehgal, B.P.S., *Human Rights in India: Problems and Prospective*, Deep and Deep Publications, New Delhi, 1996.

Sen, Geeti, *Indigenous Vision*, Sage Publications, New Delhi, 1992.

Sharma, G., *Human Rights and Social Justice*, Deep and Deep Publications, New Delhi, 1997.

Sharma, M. L., *Gandhi and Democratic Decentralization in India*, Deep and Deep Publications, New Delhi, 1987.

Shiva, Vandana et al., *Ecology and Politics of Survival*, Sage Publications, New Delhi, 1991.

Sumi Krishna, *Environmental Politics*, Sage Publications, New Delhi, 1996.

Wadhwa, M., *Gandhi Between Tradition and Modernity*, Deep and Deep Publications, New Delhi, 1991.

26

Clean Environment as a Fundamental Right

Kanchan Saxena

The deteriorating environment around us is posing a global challenge. We are facing a great crisis. World is becoming ecologically unstable, socially alienated and economically non-viable. These problems have two dimensions—ecological and socio-economic. The *bhogavada* or consumerist culture has mechanized human beings and produced things which degrade the environment. Man has lost touch with the biological realities; his love for nature is on the wane or is already gone. His attitude has become commercialized. He is not concerned with how his rights and actions affect his fellow beings, the society and the nature at large, so long as they produce immediate benefits to him.

Man is the product of five life-giving and supporting elements, i.e., earth, water, fire, air and ether. It enjoins on him to protect these elements. Instead of living in nature and using nature for his welfare in perpetuity, man started milking nature to the finish. He forgot his past and his future. He not only started living in the present but also started living in today alone and acting in the manner as if there is no tomorrow. This has led to an imbalance in nature as also in human society.

The conflict that is inherent in man's quest for happiness through material advancement has taken a variety of forms. There have been world wars, religious wars, social conflicts and psychological tensions all around. The whole human society is getting torn into pieces. But above all these men have waged an incessant war against nature endangering the very survival of life on the earth.

The roots of environmental problems lie in the style of development otherwise known as model of development. There is on the one hand growing demand for resources to meet the basic needs of food, shelter, education and health of the poor, and on the other hand and an equally growing demand for meeting the ever lasting greed as expressed by the conspicuous consumption of the rich which are having adverse effects on environment.

Much of the environmental problems we face today can be traced to the style of development being pursued in the world. Intensive use of modern science and technology to exploit the nature for more goods without being concerned about its ecological consequences has now landed us in serious environmental troubles. While our goal was to reach newer heights of happiness, we now find our future uncertain.

Technological developments ultimately made man so powerful that he started over harvesting nature. The old equilibrium was disturbed and man lost sight of his being a part of nature. The dangers inherent in such an approach to life became visible and hence now the concern about environment and ecology.

We have now come to a stage where human species benefit at the cost of other species. I do not believe that ethical principles are eternal and inherent. If they are so, they have to be modified in the light of new challenges emerging from our material prosperity. One may agree with the view that social consciousness and ethical values have not grown with the new challenges of our society. It may now be the time to do so. "We have to consider the meaning of environment in a deeper sense, the sense of our moral relationship; with the non-human world. Of course, the two senses are related in that the value of the environment is changed and considerably extended if the relationship is construed not just as an instrumental one but a moral one."[1]

It seems clear that we have a *prima facie* duty not to pollute the environment that is, we are morally obligated, in the absence of over-riding moral considerations, not to pollute. This *prima facie* duty may be understood as being based on the needs and interests of the existing human community, though one may wish to argue that it is based on the needs and interests of future generations as well. Human welfare, in fact human life, crucially depends on such necessities as breathable air, drinkable water, and eatable food. In the absence of overriding moral considerations pollution is morally unacceptable precisely because it is damaging the public welfare. On an alternative construal, the *prima facie* duty not to pollute may be understood as being based upon a basic human right, the right to a clean (livable) environment.

The law is enacted and enforced to protect the rights of individuals while the ethical values are there to ensure that each one performs one's duties towards other individuals and institutions maximally. The difference in approach, at times, brings ethical values and laws in conflict with each other probably because the source of ethics is the heart, whereas that of formal law is the head. This disjunction is the source of much of the troubles of the modern world. This conflict is more pronounced in matters of ecology and environment. Here comes the question whether right to a livable environment can properly be considered to be a human right. Is the right to a livable environment is seen as a basic and inalienable human right; this could be a valuable tool for solving some of our environmental problems, both on a national and international level. In this very context, I would like to discuss that whether we have the right to a clean environment as a fundamental human right or not? Are there any philosophical and conceptual difficulties in treating this right as a fundamental right?

I would like to begin the discussion with the view that if it is supposed to be granted that every existing human being has a right to a livable environment, it yet does not follow that every act of environmental pollution is morally unacceptable. In many cases, there may be over-riding moral considerations. For example, consider the case of a city's sole power plant. Suppose that its operation does in fact generate some small measure of air and water pollution. It is hard to think that the continued operation of this plant is morally unacceptable. Here, it seems clear that the public need for a continued energy supply is sufficient to warrant that it be continued.

It seems from the above description that many concrete cases of environmental pollution confront us with grave difficulties of weighing conflicting rights and interests. There may be many other cases of environmental pollution, which are otherwise essential for the fulfilment of our basic needs and the survival of human race on the earth. Now, the question arises: in which of these cases does the right to a livable environment demand the closing of such cases? Certainly, the general public interest in the quality of the environment always demands to be recognized.

Humanity and environment are not two discrete phenomena. They constitute an integrated lifecycle system and hence have implications for each other. Humans are part of the nature. They can manipulate nature and ignore its laws of balance only to the extent that the life sustaining capability of nature is not destroyed. "Man has to take quite a lot from the environment for his survival and development. But this taking must be like taking milk from cow. The cow's life is sacred; it must be preserved so that it continues to give milk. Such a concept puts many limits on man's use of nature."[2] Let us build dams if they are needed, but let us also see that proper habitat is to be provided to animals affected by submergence; let us put that much land under forests somewhere else at the same time we build the dams; and also let us rehabilitate the people uprooted from the project area in a manner that they are better off than before. What we have done so far in our country is not enough. We must do much more for the animal life, as well as for the poor.

Man has to learn to live in harmony with his fellow beings. But that is not enough. He has also to be in harmony with nature, animals as well as inanimate. Man should respect nature; he should take care of nature and do nothing, which is not sustainable economically as well as ecologically. Man is a part of nature. He has no existence outside nature. His life is tied to oxygen he breathes, the food he eats, the water he drinks, etc. His all primary needs find satisfaction from nature. Indira Gandhi said: "Some people still consider concern for the environment an expensive and perhaps unnecessary luxury. But, the preservation of the environment is an economic consideration since it is closely related to the depletion, restoration and increase of resources. In any policy decision and its implementation, we must balance present gains with likely damage in the not too distant future. Human ecology needs a more comprehensive approach."[3]

The scenario is fully changed today. Man has forgotten his intimate relationship with nature. Modern man armed with modern science and technology has tried to exploit nature to an extent that its life-sustaining capability is jeopardized, endangering the very survival of life on earth. If it is so, then another question may be asked whether an individual or a nation have unfettered rights to exploit nature? If an individual does not have the right to commit suicide, does he have the right to deprive others of life? Each individual on this earth has an equal right to live on earth, but he also has the duty to see that he does not destroy nature nor does he use more than what legitimately belongs to his share.

The point that emerges from the above is that humans have the inalienable right to use nature to save themselves from the jaws of death. Beyond this the right is tempered. It can be enforced only within two ethical obligations. Firstly, people do not use more than their share of the resources and secondly, they give back to nature that maintains its life supporting delicate balance. We have to move towards a society in which man-nature relationship is not based on wants but on needs; and on the unity and integrity of all life on the earth. "On the ethical plain, historically, two streams of thinking have existed. One is the oriental thinking of human beings living in harmony with nature, and the acceptance of the need for the restraining of human desires rather than incessant attempts to satisfy them."[4] Mahatma Gandhi said: "There is enough in this world to meet every one's need but not every one's greed. We can say that there is enough for the need of every nation but not enough for the greed of even one nation."[5] He further said that so long as man used nature to meet this needs, no harm was done. But, the moment he used it to satisfy his greed, he became destructive. Unfortunately, much of the modern law is based on promises, concepts of an acquisitive, permissive, and consumerist society. All these promote greed. There is, therefore, an urgent need to evolve lifestyles and development models, which are not greed promoting, they are need satisfying.

The concept of earth as the sustainer of the human race, of the race itself as one family and of the necessity to nourish and protect earth from despoliation is thus deeply rooted in our ancient thought and cultural heritage. That human beings as the children of God, as the finest and the best in God's creations hold the key through contentment and self-abnegation to the sustenance of the universe has been forcefully brought out in these words:

> Please remember that this world is the Lord's creation,
> And everything that exists here is His
> He pervades the whole universe.
> Be content with whatever is left by Him
> And do not greed the wealth of others.[6]

In this model, development acquires a new meaning. It essentially means cultural enrichment through a judicious balance between economy and ecology and between law and ethics.

Environment is the life support system of humanity. Clean, healthy and safe environment is a human right. According to Blackstone, "Every person does have a right to a livable environment, precisely because a livable environment is a necessary condition of fulfilling human capacities. The recognition of the right to a livable environment, he further argues, is now incompatible with a continued recognition of property rights, but the former will necessarily qualify or restrict the latter."[7] I think that we must recognize both the need for such restrictions and the fact that none of our rights can be realized without a livable environment. Both public welfare and equality of rights now require that natural resources not be used simply according to the whim and caprice of individuals or simply for personal profit. This is not to say that all property rights must be denied and that the state must own all productive property, as the Marxists argue. It is to insist that those rights be qualified or restricted in the light of new ecological data and in the interest of the freedom rights and welfare of all.

The answer then to the question, is the right to a clean environment a fundamental human right? is yes. Each person has this right as being human and because a clean environment is essential for one to fulfil his human capacities. And given the danger to our environment today and hence the danger to the very possibility of human existence, access to a livable environment must be conceived as a right, which imposes upon everyone a correlative moral obligation to respect.

Some philosophers further argue that the right to a clean environment ought to be granted a formal legal status. "If the right to a decent environment is to be treated as a legal right, then obviously what is required is some sort of legal framework, which gives this right a legal status."[8] A good case can be made for the view that not all moral or human rights should be legal rights and that not all moral rules should be legal rules. It may be argued that any society, which covers the whole spectrum of man's activities with legally enforceable rules, minimizes his freedom and approaches totalitarianism. There is this danger. But just as we argued that certain traditional rights and freedoms are properly restricted in order to assure the equal rights and welfare of all, so also it can plausibly be argued that the human right to a livable environment should become a legal one in order to ensure that it is properly respected. Given the magnitude of the present dangers to the environment and to the welfare of all humans, and the ingrained habits and rules, or lack of rules, which permit continued waste, pollution and destruction of our environment resources, the legalized status of the right to a clean environment seems both desirable and necessary.

Notes

1. Nicholas Low and Brendan Gleeson, *Justice Society and Nature*, p. 133.
2. R.P. Mishra, *Environmental Ethics*, p. 182.
3. Gandhi, Indira, in her inaugural address to the Non-Aligned Conference in 1983.

4. R.P. Mishra, *Environmental Ethics*, op. cit., p. 33.
5. Ibid., p. 122.
6. *Isavasopanishad.*
7. Cited in A. Thomas Mappes and Jane E. Zembaty, *Social Ethics*, p.340.
8. Ibid., p. 343.

27

A Prospective Study of Environmental Trends in India

Mukulika Hitkari

The domain of environment, in the wider sense of the term, has assumed awaringly-increasing importance in the present global context of 1990s and early twenty-first century. Being the *locus standi* of all the spheres of socio-economic development the governing object for the study of environmental trends towards a prospectively sound economic sustenance should be not a mere theoretical attempt at it but to make use of it for demonstrating its role in emerging cultural landscape or the regional diversities. In other words, it is the ethics of environment, which needs to be imperatively understood and pursued so as to overcome the problem of unprecedented devastation caused by a un-eco-friendly approach by mankind to nature.

Against this backdrop, this paper has been divided into three parts for an intensive analysis of the environmental trends. The first part deals with the desirability of environmental improvement programmes. The second part deals significantly with the environmental stress aspect being a response to environmental hazards. Finally, the concluding part assesses the role of environmental planning and management in creation of awareness for a positive approach to environmental trends in India.

Desirability of Environmental Improvement Programmes

Population growth, personnel skill and land conversions into several economic and social institutions witnessed rapid rate of shrinking of 'natural' or 'wild areas'. The 'economic man' equipped with superior technology, in lieu of satiating his indispensability has since continued to exploit the natural resources not withstanding the rebounding repercussions on his own existence. Even, the environmentalists group of conscious people have alarmingly warned the economic man against the destructive impact of unplanned, unscientific and reckless exploitation of nature.

Environment and society are factually interdependent and interrelated. On one hand, natural environment has helped in evolution of social structures, while on the other hand the existence and quality of environmental sustenance significantly rests on the responses of these social structures to the environment. Multifarious issues like diminishing environmental quality, disruption of earth's natural ecosystems creating ecological imbalance, depletion of resources, etc., have been under constant rise. However, the remedy for improvement in such programmes can be approachable if value-based judgements are implemented considerably, i.e.,

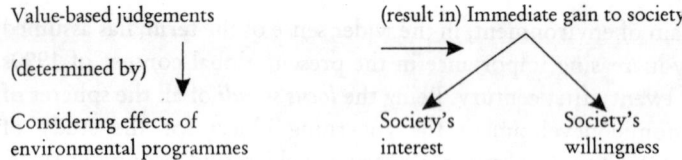

To remove the constraint of costly long-term investment eclipsing the tempo, jeopardizing public interest and tabooing implementation of improvement programmes, the requisite is stern determination to pursue environment programmes ethically. However, awakenment and public concern has reached emotional peak in society but "this spirit has to be sustained for a longer period".

Probingly, human activities have a deep impact on environment—directly participating and influencing its processes. As per the disturbing trends, the dawn of efficiently sophisticated technology-based industrial revolution in late nineteenth century initiated a hostile 'man-environment relationship' leading to reckless and indiscriminate rapacious exploitation of natural resources for urban expansion—altogether creating most of the present day 'environmental crisis' of global dimension. A step towards cautioning, even the World Watch Institute (USA) released authentic and well-documented report titled, "State of the World" in 2000 wherein it listed crucial factors to be heeded to, viz., depletion of ozone layer, greenhouse effects, soil erosion, deforestation and population explosion. It suggested immediate elaborate action plans at global scale to safeguard the earth from perilous effects of man-made environmental diseases so that future generations may not de-hold themselves in the long run.

Environmental Stress: A Response to Environmental Hazards

Environmental hazards are the processes resulting in severe environmental disasters, thereby causing environmental crisis, the intensity of disasters weighed in terms of quantum of damages done to the human society. The constraining factor being the rate of exploitation of earth's finite resources which are rising more rapidly than the rate of population growth. Undesirably causing exhaustion of certain valuable non-renewable resources. Therefore, when cumulative effects of environmental hazards and disasters become so immense that tolerance limit of natural environment is surpassed and environmental balance is disturbed, the resultant state of highly disturbed environment is called 'environmental stress'. The following chart further explains alternative parallel terms:

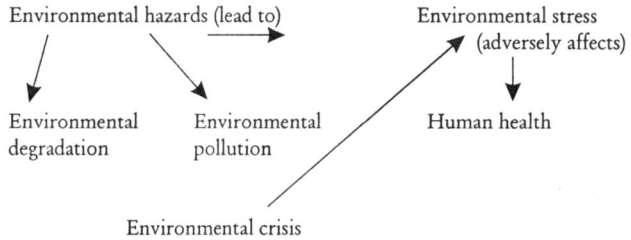

Environment degradation broadly incorporates inclusion of both natural and human factors. It can be ascribed to the following:

(a) exponential population growth;
(b) accelerated pace of scientific and technological development;
(c) ambitious projects fastening the pace of economic development;
(d) industrial expansion, sprawling urban growth and agricultural development;
(e) philosophical and religious social outlook;
(f) man-environment hostility;
(g) ignorance and lack of environmental perception;
(h) lack of public awareness towards environmental programmes;
(i) poverty;
(j) affluence and richness; and
(k) unscientific and illogical exploitation of natural resources.

It is apparent from the aforesaid discussion that the processes and causes responsible for such degradation present a future gloomy picture. Therefore, it may be evidently borne in mind that the pace of developmental work has to be adequately maintained—though not at the cost of environment—since it is the latter which ensures our existence.

Environmental pollution is restricted to the destabilization of natural components by human factors/activities alone.

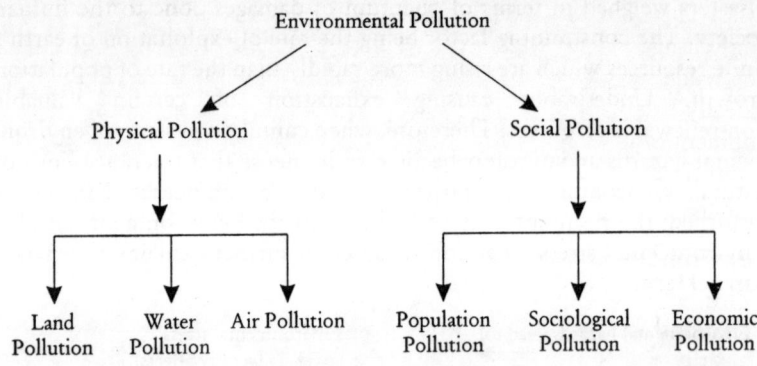

As far as physical pollution is concerned, it is inclusive of land-related problems, e.g., accelerated rate of soil erosion, desertification, soil pollution, salinization, etc. Again, water pollution relates to that of sea-water, groundwater, streams and lakes. In this context, the pathetic story of Ganga at Kanpur city clearly violates WHO norms. About 200 million litres of waste sewerage water—5.8 MLD (million litres per day) of untreated toxic effluents from 151 tanneries, industrial units and thermal power plants are discharged daily into the river—threatening its very sanctity. But, hopefully, after the completion of the ambitious Ganga Project, 75 per cent pollution would be effectively stopped. Furthermore, air pollution involves depletion of the ozone layer, increasing concentration of greenhouse gases in the atmosphere, etc., which detrimentally increase the temperature of earth's surface. Notably, USA ranks 1st in emission of greenhouse gases in developed nations whereas India ranks 2nd (next only to China) among developing nations with 250 million tones of gas emission, according to the survey report of National Environmental Research Institute, Nagpur. The level of air pollution in metropolitan cities like Kanpur has gone up—with emission from vehicles registering a contribution of 55 per cent. Means of transportation a constraining pollutant if not complying with traffic rules emits human volcanoes and disturbs the global radiation balance, resulting in chronic physiological ailments and jeopardizing human health.

Another area of social pollution relates to population growth caused by illiteracy, unawareness, conservative outlook, inadequate employment levels, etc. Social pollution includes educational and social backwardness, crimes, perpetual quarrels, wars, communal riots, etc., whereas economic pollution incorporates poverty and unemployment.

Therefore, certain suggestions should be adhered to in this regard like clean air and water, fertile land, rich biotic community, maintenance of wilderness and recreational areas, sufficient mineral resources and ideally balancing relationship with environment are essential for development of human society. Thus, 'economic man' should not be wholly converted into a 'technological man' but should remain as a 'natural man'.

Environmental Planning-cum-Management

This is essential for ground-level accessibility by man-environment relationship as well as to ensure nature's prominence in its natural clean healthy self. Efficient management is based on efficient micro-level planning hence the two corresponding to each other.

Perceptibly, five fundamental areas have to be envisaging treated as 'priority management' aspects in the global domain where environment is but a verbal and theoretical policy measure than it being a practical one. These five components are:

— environmental perception and public awareness,

— environmental education and training,

— management of resources,

— environmental impact assessment, and

— control of environmental degradation and pollution.

These will beneficially have their impact on socio-economic development of the country along with survival and stability of biosphere and ecosystems. In the era of competitive globalization, decision-makers are now becoming increasingly aware—evaluating and implementing priorities and policies. Thus, the implication is for a sound environmental planning process for enabling optimal, judicious and gainful utilization of both renewable and non-renewable resources resulting in development activities; conservation of what is rare and precious in nature; and preservation of quality of environment for the healthy growth of life.

As a concrete step in the aforesaid direction, various international conferences, seminars, agreements, declarations, symposia, etc., have been signed towards furtherance of ecological balance, biodiversity, etc., notably significant being the Stockholm, Hague and Helsinki Declarations, the 'First Earth Summit', i.e., UNCED (United Nations Conference on Environment and Development) held in 1992 at Rio De Janeiro and 'Second Earth Summit' (Plus-5 Summit) at New York in 1997. The historic Kyoto Protocol (i.e., Kyoto Thermal Treaty) holds special mention owing to the acceptance by both the developed and developing countries in principle to initiate concrete steps in checking global warming. Moreover, GEF (global environment facility) has been established with the assistance of World Bank, UNDP and UNEP.

To conclude, despite all this, what is most emphasizingly required is 'mass participation' and 'mass involvement' towards a poised eco-friendly approach as a response to the plight of the persisting environmental trends.

Undoubtedly, utilization for development is a requisite element, but it should be remembered decisively that nature should be 'explored' in all its diversities, not 'conquered'. Destruction of habitat, felling of trees, etc., be supplemented and rehabilitated with further plantations. Such an ethical viewpoint would definitely contribute in stabilizing the ecosystem.

Moreover, 'awareness' towards maintenance of a clean hygienic environment should be taken as a mission—a drive to attain the socio-economic goals and infrastructural development of the economy. Above all, the study of a novel discipline 'Environmental Tourism' should be firmly encouraged.

In a nutshell, the legacy of economic development manifests itself in environmental pollution, for which adoption of strict controlling measures are the urgent calls. Owing to substantial financial investments as well as transformations in patterns of living and energy use—social and political factors can play a major role in meeting the above goal. Though it is virtually impossible to eliminate pollution, its curtailment can itself serve a long-term purpose.

Hence, 'individualistic' approach to environmental conservation is to be discouraged as one has a tendency to acquire more than one's normal share from nature. Efforts should verily be made to create such a situation, which makes an individual responsible to a group, society or community for his activities towards the uses and misuses of environmental resources. Nevertheless, environmental planning is the best medium for achieving social justice, equitable distribution of economic wealth and uniform regional development.

References

Bahuguna, S. (1990) "Tehri: The Dam of Discontent", *Yojana*, vol. 34 (10), pp. 9-12.

Fraser, D.F. (1983) (ed.) "The Unity of Ecology", British Association for the Advancement of Science, vol. 20, pp. 297-306.

Kumar, V.K. (1982), *Kanpur City: A Study in Environmental Pollution,* Tara Book Agency, Varanasi.

Rao, K.L. (1975), *India's Water Health,* Orient Longman Limited, New Delhi.

Singh, J. and Singh, D.N. (1988) *An Introduction to Our Earth and Environment*, EDSC, Varanasi.

Singh, Savindra (2003) "Environmental Geography: Conceptual Framework," *National Geographer*, vol. 24(1), Allahabad, pp. 13-20.

28

Environmental Ethics and Its Effects on Man

Deepshikha Banerji

We are passing through a phase of environment evolution. Man was never so concerned about environment as he is now. Research in the field of environmental studies reveal that man is part and parcel of this environment. He cannot exist apart from it. Besides the four elements (air, fire, earth and water), animals, humans, trees, plants, etc., all combine to make ecology and nature balance everything one against the other. Therefore, environmental science talks of 'nature'—various forms of scientific studies, devises or techniques by which one can study nature without harming it.

If we try to define environment or ecology, it would mean a mixture of physical and biotic situations. In other words, it would mean a combination of the biosphere, lithosphere and hydrosphere as well as the atmosphere in so far as it concerns living beings. The term 'environment' means to encircle or something which is all around—in short it would mean surroundings. If we look into the meaning of the term, it quite clearly indicates that it concerns nature. Defining ecology, Charles Southwick claims that ecology is the scientific study of the relationship of living organisms with each other and with their environment.

Thus, we see that the sum of all conditions, agencies and influences, which affect the development growth life, and death of an organism species or race is considered in its true sense, namely, its concern is with environment.

Environment now is concerned with philosophy. Philosophy too is no longer abstract and non empirical, but rather it is now concerned with genuine and empirical problems which concern man. Environmental ethics is the philosophical approach to environmental studies. Philosophy gives a holistic approach and a comparative attitude to the problem of environment. By applying a rational approach to the problem, we come to the conclusion that man has grossly misused his surroundings. He has used it to his benefit without bothering to return to nature what he took from it. Nature, on the other hand, like an indulgent mother allowed man to become a spoilt child. But the child has grown up now, realizing all the harm that he has wrought. He now realizes that it is time to make up for his misdemeanor. Keeping a philosophical attitude in mind we must analyze the relation of man with nature.

We have to now examine why was not there an environmental crisis in the ancient ages. What is the reason that before us now comes the problem of environmental imbalance? Long ago, we had established a balance with nature. Philosophy at that time was religious and metaphysical and thus its attitude towards everything was religious. Thus, nature too was holy and divine, something to be held sacred, something to be respected. Man dared not transcend physical or natural laws, he was subservient to it. But with the rise of science came the empirical material and the positivist approach.

It taught man to use nature as a means to an end, to blindly destroy and harm it. Man became a marauder robbing nature of its glory. Man no longer believed nature to be divine. Material benefits mattered to man. He no longer cared for nature. He destroyed its bounty, pilfered its wealth and believed that we were progressing. Man believed that science and technology was the parameter of development. Now man has reached his limit; destruction awaits him at every corner.

For the sake of survival, early man by trial and error tried to understand and follow the norms of nature. So, it is not just the modern man who is concerned about environment and ecology. The ancient man too was concerned about it. The ecosystem of the world is extremely complex and interrelated. A lot of things coexists in it. First of all, we must know and understand environment and its ecosystem. We must understand its complexities and its ramifications. Biosphere is a large ecosystem and man has no moral right to interfere and disturb it.

Now, we come to the culmination of our paper, namely, what is the relation between man and environment? Human ecology is at present in crisis almost everywhere in the modern world because man seems to be unable to adjust to the change in the environment and the mode of life, which he has himself created. The question arises as to who is the master —man or nature. Various philosophers have suggested various alternatives.

First, determinism suggests that nature is stronger than man and second comes the possibilism view, which holds that man acts according to nature. But he can bring about alteration in it and therefore is more powerful than it. The third view is neo-determinism propounded by Griffith Taylor who believes that neither nature is the master of man nor is man the master of nature. Fourth, is voluntarism which claims that to a certain level changes are possible in nature but not beyond that. After this comes the normative approach which talks of an ideal relation between man and nature and last and the most modern approach is the ecological standpoint. Man is himself a part of the ecosystem. Man and his environment complement each other. They form a complete whole. Human endeavour is concerned with nature and he causes harm to it which again hurts and harms man himself. Excessive use and wastage of water, air, fire, earth becomes a cause of destruction. The destruction of nature is the destruction of man. Urbanization, green revolution and industrialization have led to increase of material prosperity but have also led to pollution.

Man did not realize that he and nature had a queer kind of relationship. A relationship of reaction and counter reaction, an inter-dependence with each other. The destruction of nature would also mean death for himself. Man was playing with fire. He did not realize that science was a good slave but a bad master.

To elucidate our point, I would like to quote from the original text of the sixteen principles released in UN International Conference on Human Environment, 1972.

Man has the fundamental right to freedom equality and adequate conditions of life in an environment of a quality that permits a life of dignity and well-being and he bears a solemn responsibility to protect and improve the environment for present and future generation.... The natural resources of the earth including air water land flora and fauna and specially safeguarded for the benefit of present and future generations through careful planning or management as appropriate.... The capacity of the earth to produce vital renewable resources must be maintained and wherever practicable-restored.

The above points quite clearly show that man and nature are very close to each other. The responsibility of taking care of the earth lies with man. From here arises a view called 'human ecology'. Man is a special being endowed with a special quality; that is why, he reacts in a special way with other entities in the environment. He is social, technically-minded and capable of a high profile lifestyle. That is why, in comparison to other living beings, he is capable of changing and adjusting the environment according to his needs. That is why, Yuri Lisitse writes in his work: "The prime purpose of human ecology is to find through research into interaction between man and his environment the optional conditions necessary for man to survive and develop his physical and spiritual ability." Even though man is a child of nature his ambitious temperament knows no bounds. His search for more production, material pleasure and endless benefits for himself has caused an increase of burden on the environment, the result of which is degradation of environment, natural calamities and social perversions.

Man has to establish a comfortable environment in order to be able to perform all those activities, which he deems fit to perform. But gradually man is finding it increasingly difficult to conform to his environment. Man in search of material pleasures has destroyed and harmed environment to such an extent that man is caught in his own trap. He cannot escape the environment he has made for himself. Therefore, human ecology is in danger. Man has removed himself from nature to such an extent that an artificial environment is threatening his very existence. Speaking about human ecology, a German philosopher, Humboldt writes nature is a biological unity, which has a natural philosophy of its own. In pursuit of material benefits, we have overlooked the above-mentioned idea and thus it is relevant now to take a close look at man and his changing relations to nature. Man must reformulate his connection with the environment.

From ancient times there have been an effort from philosophers to try and understand the relation between man and nature. The question arises as to who is more powerful man or nature. There are several answers to this question. Determinism tells us that man is not the controller. He is ruled by nature. He is a child of nature and therefore nature is more powerful than him. Man is limited by his own physical and mental capabilities which cannot transcend his body or his mind. As early as 420 years before Christ Hippocreatus talked of air, water and place as something, which caused human beings to benefit, Aristotle wrote that people of a colder climate were hard working while those of warmer climate were more philosophical but they lack inspiration to work hard. Bodin in the sixteenth century and later Montesque of the seventeenth century were also talking in the same lines discriminating between the working capacity of people in colder climates and the people of warmer climates.

In the modern age, many philosophers have defended determinism. E.C. Sample, Darwin, Humboldt all felt that even though man thought that he was the master of his own destiny it was really nature, which held his destiny in its hands. But actually this view is erroneous; man has mastered certain natural spheres. He has mastered distances and has made the world a small place. He has learned to cure diseases. He has learned to save himself from heat and cold in so many ways. He has overcome nature. His rationality, knowledge, science and technical education has taught him to control environment for personal benefit.

Possibilism is another alternative. This theory believes that even though man is governed by nature yet man has the power to alter it to suit his purpose. Fabbere, Carl Saver, etc., believe that there is an adjustment between man and nature. Man is not a slave of the environment. He uses nature to fulfil his needs. He conquers natural forces and converts them to make them beneficial for him. But this view too is unjustified. It tries to show that man can fully conquer nature. But such a view is too pompous—we cannot be the masters of something, which is the cause of human existence itself. Neo-determinism was a theory propounded by Griffith Taylor who talked of natural environment and human endeavours. Man by his knowledge and ingenuity changes nature and its variety to suit his purpose. Human activity is much affected by natural environment. Man either changes himself according to nature or changes nature to suit himself. The view that Taylor held proclaimed that neither man was the master of the nature nor was nature the master of man. Both have a functional relation with each other. For the sake of progress, man and nature should unite; they should share a give and take relationship. There must be an inter-action between them. Man can increase or decrease the speed of evolution but he cannot interfere with the direction of natural progress and evolution. Neither can he break or change the norms of nature. Actually this notion is the basis of the relation between man and environment that is concerned with scientific progress.

Voluntarism is the view held by some philosophers that nature can be altered only up to a particular level. By using various techniques, man can make use of nature in order to use it for his personal benefit. Man's development and progress is due to his way of handling nature and making use of it for material comforts. But man's handling of nature is not so perfect there are a lot of things which are faulty—one of them is pollution. Man's faulty use of nature has led to large-scale pollution. But man being rational has tried to create methods to check pollution. But so far man has not been as successful as he should have been. Man has to establish the right adjustment with his natural environment so that there can be progress and development without any harm to nature. There must be a limit to which man can exploit nature. If man does not control himself, nature will ultimately destroy him.

Ecological approach is the latest theory in this chain and it is perhaps the most easily acceptable one. I hope to defend this view. This view is of the opinion that man is a part of environment itself. Actually the environment is a triangular system with man on one point environment is on the other point and third point is that of progress, namely, economic social and cultural. Man and environment complement each other. Man's action originates from his experience of nature and his actions alter nature and affect it. But the effect is mostly harmful for human beings. Excessive use of water, land, air is causing pollution, which seems to lead man to his ultimate downfall. Urbanization and industrialization have destroyed the fragile balance of nature. Actually, man and his relation with nature are dependent upon his interaction with nature.

This view also discusses what would be the ideal situation between man and nature. It is quite obvious that population is responsible for pollution to a large extent. That is because man and his kind have increased in numbers to such an extent that nature is no longer capable of feeding them and fulfilling their needs. Man and environment must have a proper relation. There must be the right balance between the two; only them can man actually progress. The view that either man is the master of nature or nature is the master of man are both one sided views. Man by his blind, selfish, greedy and narrow views hurts and harms nature. By doing this he only brings about his own ruin; he not only invites his own destruction but also the destruction of everything else. By the increase of factories and other infrastructures involving progress, is fast making nature into an enemy of man. There has, therefore, arisen the necessity of bringing man and environment close to each other rather than making them enemies of each other. To keep the ecology in order it is necessary to maintain and keep ecological balance in order. The great sage Charak had said: "For a healthy life pure water pure earth and fresh air were most required."

Thus, we come to the conclusion that in the present age we have come to a situation where things are very grim and there is an environmental crisis. The real reason is our materialistic vision, experimentalist psyche and consumerist mentality. Man is himself responsible for this situation. Therefore, it is up to him to find a solution. The blind forces of nature have been hurt and harmed by man's selfish and careless desire for personal gains. Scientific progress has misguided us; instead of caring for nature we have misused it. We have learnt to care for material wealth alone—anything, which has ethical, or non-physical worth is not valuable for us. We are not careful in protecting our resources. Von Humboldt believes that nature is not just matter, it is something spiritual. Matter is not something, which is lifeless; it is the power inherent in nature. Ecological philosophers believe that there is a unity in nature. They are searching for that unifying force which binds everything together. Only when our own existence came under a cloud and death threatened us at every corner did we turn to look at nature. Man now has decided to provide the healing touch. If we look into the problem closely, we can come up with the following solutions: First, we must have a gentle approach towards nature. Even if there are any technological advances, they must be made in such a manner that they do not destroy the ecological balance. Second, there must be compulsory education regarding the use and conservation of resources. Third, there must be a change of outlook—we must give up the material aspect and think in terms of the metaphysical and spiritual element in nature. Finally, we can sum up by saying that science must be more humanistic.

29

Moral Status of Non-Human World

Mritunjay Kumar and Purnendu Shekhar

Though philosophers have differences of opinion over the modes of existence of world, they all agree that there is a world, which has got immense practical value. We may classify the things of the world into two categories, namely, the living or biotic component and the non-living or abiotic component. The biotic component includes all types of living organisms (plants and animals). The a biotic component includes the non-living materials (soil, water, air, etc) and the forces of nature (light gravity and molecular energy). The living organisms exist in an environmental setting of which they are a part. The environment influences every aspect of life and the activities of organisms affect their environment. In this way, there is an interrelationship between biotic and abiotic components as well as the relationships among the individuals of the biotic component. The study of interrelationship of such nature is called 'ecology'.

For philosophic reason, let us classify the living organism into two important classes—human and non-human. Animals, birds and vegetation constitute the non-human world. They have lives and they emanate moral values. Human world indicates men's living in society, in the west, has immense ethical import. Morality, in the west, is regarded as a social enterprise. Human beings who live in society and perform voluntary actions but exercising their freedom of will are the only moral beings. Those who do not live in society and lead a life of sage, their actions cannot have moral worth. Morality also has got nothing to do with animal's conduct. In other words, actions of non-moral beings do not account for moral considerations. Their actions are considered non-moral, as they do not have reason or rational faculty. But does it mean that non-human beings be used as means for human interest? Should human beings be unkind to their existence and their right to live?

In recent times, many thinkers including moral philosophers have woken up from their deep slumber to realize the moral values of non-human beings. They have felt the urge to extend the domain of morality so as to include non-human beings also. It is high time men realized the ethical values of animals and plants. Ecological system has already established how human life is connected with animals and plant. Harming one form of life upsets the entire chain of lives. For our moral consideration, let us begin with vegetation. Vegetation indicates green cover spread over our planet. Forest is an important component of the vegetation. Hanson defines forest as "a stand of trees growing close together with associated plants of various kinds". Forest is the most important natural habitat for wildlife. Of late, man has destroyed the forests to an alarming extent. As a consequence, the wildlife has faced extinction, the water table slides down and down, clouds are no longer attracted to rain in those particular areas and children simply are deprived of picnic-spots. Gradually, such areas will turn into desert in the long run. It is believed that during King Akbar's period, Rajasthan did have a dense forest. Akbar used to enjoy hunting in the forest where presently the Thar desert has come up.

In Himachal Pradesh destruction of forest has caused the land sliding like menace. If the trend of destruction of flora and fauna continues, the mountain ranges will no longer catch the fancy of the tourists. We will have to see that the sense of beauty of the nature over the mountains provides us a paradise on the earth. The great poet, Robert Frost once got so bewitched by the beauty of forest that he stopped his horse in the middle of the forest. To quote him:

> My little horse must think it queer,
> To stop without a farm house near,
> Between the woods and Frozen lakes,
> The darkest evening of the year.
>
> The woods are lovely dark and deep,
> But I have promises to keep,

And miles to go before I sleep,
And miles to go before I sleep. *

The way, the jungles are being denuded in India, it appears that coming generations will only hear about it in the storybook. Man must realize that jungles sustain their lives. Our existence will be endangered if plants do not support it. Man boasts of beings moral. But he cannot prepare his own food, the way plants do through photosynthesis. Plants provide sustenance to our lives – is this fact not enough to adduce values to the vegetable world? In order to consolidate this fact it is not out of place to highlight the importance of forests.

Forest is one of the most important natural habitats for wildlife. Man for commercial and recreational purposes utilizes it. Many herbivores find shelter and carnivores search their prey in the forest. Many wildlife store food supplies and breed in the forest. Green plants of the forest are food-producing organisms and are primary producers for the 'food chain'. They trap energy from the sun and use it to transform CO_2 from the air, together with water and nutrients from the soil into food substance like starch, sugar, through the process of photosynthesis. These foods are stored in the form of fruits, nuts, seeds, nectar and wood. Therefore, forest serves as an energy reservoir, trapping energy from sunlight and storing it in the form of biochemical product.

Forest plays a key role in purifying the atmosphere by consuming CO_2 and releasing O_2. So, removal of plants and trees would disturb the composition of natural air. An acre of forest absorbs 4 tonnes of carbonic acid gas and recycles 8 tonnes of oxygen into environment.

Forest is instrumental in soil conservation and keeping the water table intact. In natural forests, the trees roots bind the soil and about 90 per cent of the water falling on the forests is retained either in humus or in the plants tissue. The forest thus acts as a soaking device and plays a vital role in the hydrological cycle. The rainwater thus soaked up is gradually released over the days and weeks, which supply to streams and rivers during dry seasons. Hence, it is important to retain forest cover in upland catchments areas in order to prevent flooding and soil erosion. It has been estimated that in India 60,000 million tones of topsoil was carried away annually by rainwater from deforested areas. The washed away topsoil silts riverbeds, reservoirs to reduce their holding capacity and flood the surrounding areas as natural calamity.

---

* Palgrave's *Golden Treasury (An Anthology)* Robert Frost's poem "Stopping by Woods on a Snowy Evening".

Nowadays, the tendency of deforestation is increasing day by day. Man is cutting forests and gradually the wildlife is disappearing. wildlife is defined as "the uncultivated flora and the undomesticated fauna amongst the plants and animals". In the modern sense, according to Mahajan (1961), wildlife means life in any form (plant/animal) existing in natural surroundings. Wildlife provides recreational and economic benefits to man.

It is absolutely necessary to protect and conserve all forms of life on this planet as they are interdependent and form a chain. Nature has created them in such a balanced manner that if one form of life is disturbed it affects all the other forms of life too. The very existence of man depends upon the survival of other forms of life-both plants and animals. So, the destiny of humanity depends upon the survival of other forms of life.

Today, wildlife species are gradually disappearing and their number is diminishing. Many species of wildlife have become extinct or are in the way of extinction. As regards extinction of some important wildlife species, human's cruel acts are quite responsible. We kill wild animals and birds for fun, money and taste. Their meats satisfy our palate and skins are used in making purse, dresses and like items. Items made up of their skins are very costly. In western countries, there are crocodiles parks where these species are reared and brought up only to be killed later for their skins which have immense material value but unfortunately abysmal moral import.

Scientific medical researches have also led to the killings of animals to a great extent. Large numbers of chimpanzees are killed nowadays for research. Chimpanzees are difficult to breed in captivity. The standard method was to shoot a female with an infant by her side. The infant was then captured and shipped to Europe and the US from the jungles of Africa. It is estimated that for every infant who reached its destination alive, six chimpanzees are killed. Although chimpanzees have been placed on a threatened list, and this trade has now been banned, illegal killing and trading of chimpanzees, and of gorillas and orangutans, still continues. Despite an official moratorium, the whaling industry slaughters thousands of whales annually in the name of research. Dolphins are also caught and killed for human interest. Therefore, dolphins now have become a rare species. According to United States Department of Agriculture figures, approximately 1,40,000 dogs and 42,000 cats die in laboratories in the US each year.

Philosophers of the modern era seem to be evaluating the moral status of animals and birds in view of their ruthless killing and exploitation for human interest. If we follow the western line of thinking it would appear morally wrong to kill non-human beings for the sake of fun, money, research or the likes. Moral arguments advanced in this connection are that one has no right to deprive non-human beings from the pleasure of existence. It is often said that the case against animal killing is weak since animals are not rational or self-conscious being. Also that the killing of animals hardly cause suffering to other animals, and the killing of one animal makes possible its replacement by another who could not otherwise have lived. But western scholars would raise the objections that some species of apes, dolphins, whales, dogs and cats do have appreciable degrees of rational behaviours. Chimpanzees are appreciated as one of the most intelligent species among apes. They can pick up sign languages and civilized conducts if properly trained. They just lack in communicating in human language.

These species do have innate ways of communicating to one another. Whales bring out melodious songs, dolphins make buzzes and whistles, dog's bark and birds sing to communicate with their companions. Honeybees while returning to the hive perform a particular type of dance from which other bees learn the distance and direction of the food sources from which the bee has come.

It has been surveyed that the behaviour of animals points to the conclusion that they possess both memory of the past and expectations about the future, and that they are self-aware, that they form intentions and act on them. From the above description it appears that non-human beings too are persons who have intelligence and self-consciousness. Therefore, if human life does have special value or claim to be protected, the non-human beings too have equal rights to exist.

Again, it is also argued that many modes of killing used on animals do not inflict an instantaneous death, so there is pain in the process of dying. There is also the effect of the death of one animal on his mate or other members of the animals social group. There are many species of birds in which the bond between male and female lasts for a lifetime. The death of one member of this pair causes distress, and a sense of loss and sorrow for the survivor. The mother-child relationship in mammals can be a source of intense suffering if either is killed or taken away.

All there factors suggest that the life of animals and birds is no less important. It is easy to fathom moral values existing in the non-human world. But man as a moral being should learn to honour the life of non-human beings.

Morality is also based on the principle of equality it implies the sense of equal respect for all kinds of life on the earth. Hindu views of morality can be cited in this connection to consolidate our assertions that all lives on this planet must be honoured as being part of the Supreme Divinity. Also human beings should accept the principles of non-violence, coexistence and acceptance. Hindu morality believes that all lives are but sparks of the Supreme Divinity. If we cause harm to any life it ultimately pains Him. Man cannot give life, so he has no right to take it away. Hindu moral wisdom preaches the spirit of "live and let live". One must always remember that the God is the only Moral-Governor of this universe. He maintains the cosmic order. The principle of 'rita' enshrined in the *Vedas* confirms this fact. God is the only creator, sustainer and destroyer of this cosmos. Men, therefore, must nurture a wide cosmic-vision for the sake of ultimate morality so that all forms of life on this planet are honoured and protected.

Ancient Hindu scriptures express concern for all living creatures including plants. Plant worship, since time immemorial, has been a part of Hindu tradition. We still worship neem, tulsi, peepal, banyan and like plants. Our epic talks about *gurukul* system where teachers used to impart education to their pupils amidst natural surroundings of forest. Right from the childhood, students were taught to respect nature and animal kingdom.

Modern science too has proved that plants do have a sense of feeling, sorrow and pleasure. Indian scientist J.C. Bose displayed this fact on 'Cresco graph'. When electric shocks were given to plants, they wreathed in pain—the monitor of Cresco graph displayed their pains through curve linings.

Man slaughters animals and destroys vegetation. From moral points of view, it is indeed an immoral act to kill animals and birds for satisfying our palate. If man is given the similar treatment, just imagine the agony and pain he will have to undergo. Man, therefore, must understand the values of all lives. We have already discussed how nature has arranged and balanced all sorts of lives. If one form of life is disturbed, it upsets the entire gamut. It is, thus, necessary to accept the moral status of animals, birds and plants. Ecology has established this fact. Now, it is the turn of philosophy to recognize and incorporate this fact (moral status of non-human world) for moral consideration. Moral concern, therefore, should have all-inclusive ranges so as to include behaviours of animals and birds. Simultaneously, moral philosophers should show total awareness towards flora and fauna as they influence human life to a great extent. Philosophical moral wisdom, therefore, should widen its horizon to encompass all forms of lives on this planet. The great moral wisdom has long been an integral part of Hindu tradition, but the same is yet to be incorporated in the western concept of morality, which considers morality as social enterprise only. But, very recently, they have come forward to restore and respect the rights and interests of non-human world. Further, it will be a matter of great interest to peep into the exclusive zone of animal kingdom's morality.

References

Palgrave's *Golden Treasury (An Anthology)* Robert Frost's poem, "Stopping by Woods on a Snowy Evening".

Shukla, R.S. and Chandel, P.S., *Plant Ecology*, S. Chand & Company, New Delhi, 1998.

Singer, Peter, *Practical Ethics*, Cambridge University Press (1993 edition), 2000.

Contributors

S. A. Shaida, Retired Professor of Philosophy, Department of Humanities and Social Sciences, I.I.T., Kanpur.
Residence: 34, Kazmi Street, Rani Mandi, Allahabad.

Rajendra Prasad, Retired Professor of Philosophy, Department of Humanities and Social Sciences, I.I.T., Kanpur.
Residence: Opposite Stadium, Rajendra Nagar, Patna.

Bijoy H. Boruah, Professor of Philosophy, Department of Humanities and Social Sciences, I.I.T., Kanpur.
E-mail: boruah@iitk.ac.in

S. K. Pal, Department of Philosophy, University of Burdwan, P.O. Rajbati, Burdwan.
Residence: F-2, TARABAG, P.O. Rajbati, Burdwan.

Ramdas Sirkar, Department of Philosophy, Vidyasagar University, Midnapore.

Bibhu Prasan Patra, Associate Professor of Business Ethics, Xavier Institute of Management, Bhubneswar.

A. Raghuramaraju, Reader, Department of Philosophy, Central University, Hyderabad.

R. C. Sinha, Professor of Philosophy, Patna University, Patna.

D. C. Srivastava, Reader and Head, Department of Philosophy, Christ Church College, Kanpur.

Residence: 117/Q/10, L.I.C. Colony, Sharda Nagar, Kanpur.
E-mail: dcsri@rediffmail.com

Richard Howell, General President, Evangelical Fellowship of India, 805/92, Deepali, Nehru Place, New Delhi.
E-mail: efiindia@del2.vsnl.net.in

P. K. Bajpai, Department of Zoology, D.A.V. College, Kanpur.

Nitish Dubey, Reader, Department of Philosophy, D.A.V. College, Kanpur.

Ranjay Pratap Singh, Reader, Department of Philosophy, D.A.V. College, Kanpur.

Sanjay Kr. Shukla, Department of Philosophy, Ewing Christian College, Allahabad.

Shiv Bhanu Singh, Reader and Head, Department of Philosophy, Ewing Christian College, Allahabad.

Rajjan Kumar, Reader, Department of Applied Philosophy, M.J.P. Rohilkhand University, Bareilly.

Avinash Kumar Srivastava, Reader and Head, Department of Philosophy, Nalanda College, Biharsharif, Nalanda.
Residence: Amber Chauraha, Beside Mahadeo Temple, Biharsharif, Nalanda.

P.K. Pandey, Professor of English, Banaras Hindu University, Varanasi.

Sanjay Kumar, Reader, Department of English, Banaras Hindu University, Varanasi.
E-mail: sank45@satyam.net.in

Sujata Chaturvedi, Department of Hindi, Christ Church College, Kanpur.
E-mail: suchaa1965@yahoo.co.in

P.K. Rath, Reader, Department of Physics, University of Lucknow, Lucknow.
Residence: 19, Reader's Flats, Faizabad Road, Lucknow.

Vishnu Ratna, Lake View Building, II Floor, 112/1-C, Benajhabar Road, Kanpur.
E-mail: ncs-india@satyam.net.in

Jyotsana Lal, Department of Chemistry, Christ Church College, Kanpur.

A.K. Sharma, Professor of Sociology, Department of Humanities and Social Sciences, I.I.T. Kanpur.
E-mail: arunk@iitk.ac.in

P. Vigneswara, Research Fellow in Sociology, Department of Humanities and Social Sciences, I.I.T., Kanpur.

Vandana Asthana, Reader and Head, Department of Political Science, Christ Church College, Kanpur.

Ashutosh Saxena, Reader, Department of Political Science, Christ Church College, Kanpur.
E-mail: asaxena2k3@rediffmail.com

Kanchan Saxena, Reader, Department of Philosophy, University of Lucknow, Lucknow.

Mukulika Hitkari, Department of Economics, Dayanand Girls' College, Kanpur.
E-mail: mhitkari29@yahoo.com

Deepshikha Banerji, Reader and Head, Department of Philosophy, Jagat Taran Girls' College, Allahabad.

Mrityunjay Kumar, University Department of Philosophy, T.M. Bhagalpur University, Bhagalpur.

Purnendu Shekhar, I. B. Road, H. Kharagpur, Munger.

Index